Magnetism: A Very Short Introduction

VERY SHORT INTRODUCTIONS are for anyone wanting a stimulating and accessible way into a new subject. They are written by experts, and have been translated into more than 45 different languages.

The series began in 1995, and now covers a wide variety of topics in every discipline. The VSI library now contains over 500 volumes—a Very Short Introduction to everything from Psychology and Philosophy of Science to American History and Relativity—and continues to grow in every subject area.

Titles in the series include the following:

Stephen Blundell

MAGNETISM

A Very Short Introduction

OXFORD
UNIVERSITY PRESS

Great Clarendon Street, Oxford, OX2 6DP,
United Kingdom

Oxford University Press is a department of the University of Oxford.
If furthers the University's objective of excellence in research, scholarship,
and education by publishing worldwide. Oxford is a registered trade mark of
Oxford University Press in the UK and in certain other countries

First Edition published 2012

British Library Cataloguing in Publication Data

Data available

Library of Congress Cataloguing in Publication Data

Data available

ISBN 978-0-19-960120-2

Printed and bound by
CPI Group (UK) Ltd, Croydon, CR0 4YY

To Paul and Jen Riddington

Contents

Acknowledgements

I am grateful to many friends, students, and colleagues for enjoyable discussions about different aspects of magnetism through which I have learned much. I would particularly like to mention Steve Bramwell, for many interesting conversations about spin ice; Andy Gosler, who told me about the bar-tailed godwit; colleagues and students at various international schools on magnetism, who have stimulated my thinking; and members of my research group and my research collaborators who have invariably been a fund of useful insights. I would like to record my deepest thanks to Katherine Blundell and Latha Menon for numerous helpful comments on the manuscript.

Oxford, March 2012

List of illustrations

Chapter 1
Mysterious attraction?

What is that mysterious force that pulls one magnet towards another, yet seems to operate through empty space? The 19th-century scientist Michael Faraday carried out many investigations into the behaviour of magnets, and the result of one of his experiments (reinacted by the author) is shown in Figure 1. It is a demonstration that many people have performed themselves when very young. A bar magnet is placed under a sheet

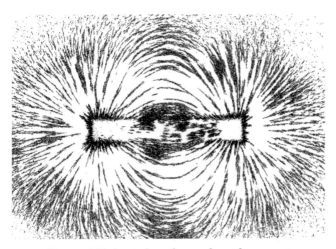

1. Iron filings sprinkled on a piece of paper above a bar magnet

of paper and iron filings are liberally sprinkled on the top surface of the paper. The filings arrange themselves in a pattern which shows how the influence of the bar magnet is felt through space. Despite the piece of paper in the way, some kind of message is propagated through the paper and causes the filings to line up. The pattern produced seems to suggest that lines of influence stream out of one end of the magnet, loop round through space, and then re-enter the other end of the magnet. For centuries, people have been fascinated by this unusual property of Nature and have asked themselves what makes magnets work. They have then wondered what can you do with them.

The struggle to understand magnetism has been long and tortuous. At various times, magnets have been claimed to be useful in detecting adultery, healing the sick, and also unlocking the secrets of the life force of the Universe. The first genuine and productive application of magnets arose in the field of navigation, where small bar magnets have been used for well over a thousand years following the invention in China of the magnetic compass. Today, magnets are employed in applications as diverse as the storage of information and the generation of electrical power, not to mention their crucial role keeping various plastic letters of the alphabet attached to the doors of refrigerators.

As we shall see in this book, the origin of the magnetism of certain rocks was debated by the Ancient Greeks and in medieval times, but it was William Gilbert who was the first to perform a series of systematic experiments on magnetism, correctly realizing that the Earth itself behaved like a giant magnet. This set the pattern for later discoveries, through Oersted, Ampère, Faraday, Maxwell, Einstein, and others. I will also describe how the elementary characteristics of atomic magnets were elucidated by the early proponents of quantum theory. Today, we have gained much understanding of processes in the Universe through studying magnetic fields in space, while closer to home, magnetic confinement explains the aurorae and offers a route to produce fusion reactors on Earth. The search continues for magnetic monopoles, and new ways of understanding magnetic materials are revolutionizing the way we now store information. This is a story underpinning the technology of the modern world, but it is also a story of a quest to understand. This intellectual crusade can be distilled into asking a simple question: what makes magnets attract things?

The power to enchant

Sometimes the hidden attraction of an unseen force can be distinctly sinister. The Sirens were bird-like women in Greek mythology whose melodious and seductive singing lured sailors to shipwreck on the rocks. In George Lucas's *Star Wars*, it was an invisible 'tractor beam' that sucked the Millenium Falcon out of its planned trajectory and dragged the unwilling spacecraft towards the Death Star. It is perhaps not surprising that the unseen power of a magnet, attracting distant iron objects through thin air, carries with it an atmosphere of mystery and menace. Is the magnet alive? Does it have a soul? Is it evil?

The discovery of magnetism began with a type of rock. Magnetite is a mineral with chemical formula Fe_3O_4. It is commonly found in various locations around the world, although it gets its name from Magnesia, a region of central Greece (the names of the chemical elements magnesium and manganese, neither of which have anything much to do with magnetite, also derive from Magnesia). Many pieces of magnetite are naturally magnetized, probably due to lightning strikes, and will therefore pick up bits of iron. This mysterious property of what became known as magnets was known by the Greeks (it is mentioned by Thales of Miletus in the 6th century BCE) and also to the Chinese (there is a reference to magnetism in literature of the 4th century BCE).

The poet Lucretius composed his famous *De Rerum Natura* ('On the Nature of Things') in the 1st century BCE. Writing in Latin hexameter, he attempted to explain Epicurean philosophy and science to a Roman audience, and he includes an extended passage on magnets. Lucretius was fascinated by an experiment he must have witnessed in which the power of a magnet was shown to support a chain of iron rings. He marvelled at how the iron rings clung tenaciously to each other and to the magnet from which they were suspended, held only by this incomprehensible force while 'swaying gently in the breeze'. He advanced an explanation of this phenomenon on the basis of the atomic theory of Democritus and Leucippus. This theory implied that everything is composed of atoms, each of which reflected the essence of the object in question. Lucretius noticed that an object's essence seems to permeate away from it through the air. Odours from food, heat from the Sun, sweat from the body, 'a damp taste of salt enters our mouths when we walk by the sea', all are examples of the same principle, that 'nothing exists without a porous texture' and all demonstrating that atoms of one substance can pass right through other substances, though often to differing degrees. These ideas form the basis of his explanation of the 'magnetic stone'.

Magnetism

> Firstly, there must needs flow out from this stone
> A multitude of atoms, like a stream
> That strikes and cleaves asunder all the air
> That lies beneath the iron and the stone.
> Now when this space is emptied, and a large
> Tract in the middle is left void, at once
> The atoms of the iron gliding forward
> Fall in a mass into the vacuum.

Thus he conjectured that iron atoms flow out of the stone, cut the air, and create a vacuum which then sucks other objects in particular directions. This is a propelling force which acts 'as a wind drives sails and ships'. Lucretius' explanation of magnetic attraction is completely wrong, but is ingenious and demonstrates serious and thoughtful engagement with this phenomenon in the physical world at a level that would not be bettered for centuries.

At roughly the same time in China, investigations of magnetism were also being carried out. There is evidence that a magnetized ladle was used for divination, testing the future by seeing which way the magnetic ladle lined up when spun. From this, it is likely that the magnetic compass was invented accidentally, and there are records in Chinese literature of the 11th and early 12th centuries in which a magnetic needle, floating in water, was used for navigational purposes. From China, the new technology drifted to Europe, where it appeared in the late 12th and early 13th centuries. By the end of the 13th century, the dry compass had been invented, a design in which the magnetized needle no longer floated on water but pivoted around on a fixed pin.

In the successive centuries, magnetized rocks were often referred to as *lodestones*, from a Middle English word 'lode' (which can be traced back as far as Beowulf) that means 'way' or 'course' and shows that the navigational use of these stones had become firmly entrenched in the imagination. A related word appearing about the same period is *lodestar*, a celestial beacon which directs your path. It often referred to Polaris, the pole star, and was imagined as a guiding star on which one's hopes were fixed and by which ones's destiny was determined. Orientation by the stars or by magnetism were crucial skills in a world without reliable maps, let alone satellite navigation. Far from luring unsuspecting mariners onto the rocks, as the Sirens had done, the magnetic compass allowed sailors to navigate repeatedly and reliably around regions of danger. But the mechanism behind the operation of the lodestone remained magical and mysterious. Given that lodestones shared with lodestars the power to guide your path, it was not surprising that many took the seemingly obvious hint that magnets possessed some kind of cosmic significance.

Magnetic healing

If magnets have a strange and mysterious power, then it might
seem entirely possible that they can intervene directly in human
affairs, in particular with mankind's unending struggle with
afflictions and disease. Indeed, the power of lodestones to treat all
manner of illnesses has been asserted for centuries. The purported
therapeutic nature of coming into contact with magnets or
ingesting quantities of magnetic rock was frequently touted.
However, before being too ready to criticize such quack therapies,
it should be remembered that much of the conventional medicine
throughout most of human history was hardly any better.
Treatments were barbaric and frequently misguided, the cure
often causing far more damage than the disease. Surgery was
carried out without anaesthetic, and remedies such as tincture of
arsenic and purging of the blood were common. From this
standpoint, magnetic therapy was comparitively risk-free. It was
therefore, unsurprisingly, popular and those who practised it were
not short of customers.

Many examples of magnetic healers could be provided, but
probably the most notorious was the 18th-century German
physician Franz Mesmer who built a cult following on his new
form of treatment based on magnetism. Mesmer was a
charismatic figure, and his view of the road back to health was one
that appeared straightforward and clear-sighted. His idea was
quite simple. Many complaints were due to the internal passage of
magnetic fluid becoming blocked by an obstruction and, once this
blockage was dealt with, the natural flows could be restored and
full health would return. Mesmer could achieve this unblocking,
and patients described the release that accompanied this magnetic
unclogging as if a warm wind had passed through them. Mesmer
practised in Vienna and then moved to Paris in 1777, and word of
the miraculous effect of his treatment spread. Mesmer's treatment
initially involved the use of magnets which could steer the body's
natural 'animal magnetism', but he eventually abandoned these

and concentrated on more elaborate rituals, staring into the patient's eyes, passing his hands over their body, touching them with an iron rod, or holding their hands for long periods of time, and eventually inducing a brief convulsion in which the offending obstruction would pass.

Mesmer claimed that animal magnetism:

> is a universally spread fluid; it is the means of a mutual influence between celestial bodies, the earth, and living bodies; it is continuous so as not to permit any vacuum; it is incomparably subtle; it is capable of receiving, spreading, and communicating all the sensations of movement.

He claimed that:

> one recognizes particularly in the human body, properties similar to those of the magnet. One distinguishes two diverse and opposed poles. The action and property of animal magnetism may be transmitted from one body to another, animate and inanimate: this action operates from a distance, without the help of any intermediary body; it is increased when reflected by mirrors, communicated, spread, and increased by sound; this property may be accumulated, concentrated, transported.

In particular, he asserted that animal magnetism:

> may itself cure nervous disorders and be a medium for curing others; it improves the action of medications; it induces and guides crises in such a way that disorders can be understood and mastered. In this way, the Physician knows the state of health of each individual and determines with certainty the origin, nature, and progress of even the most complicated of diseases; he prevents their spread and reaches a cure without ever exposing the patients to

dangerous effects or unfortunate consequences, regardless of age, temperament and sex.

Quite a claim.

Mesmer went down a storm in Paris, and very soon a number of other magnetic healers were springing up to get in on the act and take advantage of the lucrative new trade. Mesmer's livelihood was under threat, and he protested that only he had the special gift and power. To sort out the dispute between Mesmer and his imitators, as well as to see if the whole practice could be established on legitimate scientific basis, the French Academy of Sciences instigated a high-level commission to examine the whole matter. One particular member of the commission was one of the most respected scientific figures of the day who had also courted controversy by his extraordinary power to tame thunderstorms and thereby save cities. His name was Benjamin Franklin, the inventor of the lightning conductor.

Benjamin Franklin

One of the great rationalists of the 18th century, Benjamin Franklin was born in Boston, Massachusetts, in 1706, just over a decade after the Salem witch trials, a horrific example of the consequences of human irrationality. Aged 17, Franklin arrived in Philadelphia and worked his way up to a career as a journalist and author via a spell as a typesetter. He was to be later celebrated as a founding father, ambassador, politician, and statesman, but his subsequent reputation was strongly enhanced by his achievements as a scientist. Franklin invented the lightning conductor, bifocal spectacles, and the urinary catheter, developed the Franklin stove, and made important contributions to evaporative cooling, oceanography, and thermodynamics. He is perhaps most famous for flying a kite during a thunderstorm in 1752, thereby demonstrating the electrical nature of lightning. Franklin's interest in electricity grew in the 1740s as the subject became very much in vogue with the many amateur scientists who lived in Philadelphia and who formed Franklin's circle of friends. In London, the amateur scientist Stephen Gray had devised an experiment called the 'Dangling Boy' in which a young boy would be suspended from the ceiling by silk cords which served to insulate him electrically. The boy would be charged up by touching him with a rubbed glass rod. An audience could then be endlessly amused by watching the boy's hair stand on end, seeing small objects being stuck to him or even sparks drawn from his nose and fingers. Electricity was clearly the stuff not only of science but of theatre. Franklin probably saw such an electrical demonstration first in 1743 when Philadelphia was visited by an itinerant lecturer from Edinburgh, and from that moment Franklin was hooked.

In December 1750, in his Philadelphia home, Franklin was already deploying two Leyden jars, the batteries of his day (see Chapter 3), to help him slaughter his Christmas turkey by electrocution and invited his friends round to enjoy the fun. Something went wrong and Franklin accidentally managed to connect himself across the terminals, causing an almighty flash and resulting in all the blood draining from one of his hands. Undeterred, Franklin viewed this as all part of the learning experience. He beavered away at his experimental work and was able to show that Leyden jars stored their electrical charge on the surface of the glass, not in the water they contained. He was also the first to introduce the idea of positive and negative charge, deducing that electrifiable bodies either had an excess or deficit of charge, much in the same way a bank account could be in credit or deficit.

A familiar party trick, that of producing a musical note from a wine glass by rubbing a moistened finger slowly round its rim, was also popular at the time. Franklin heard a performance of Handel's *Water Music* on a set of tuned wine glasses and this provided the motivation for his invention of a new musical instrument which ingeniously automated this process. A set of glass disks, each fabricated to a specific size, and mounted on a horizontal spindle, could be jointly rotated about a single axis. The player could keep their moistened finger stationary on a single selected disk, changing note by selecting a different disk. The instrument was christened an 'armonica' and produced a tone which Franklin described as being 'incomparably sweet'. Mozart wrote an armonica composition in honour of one celebrated armonica soloist, accompanying her on viola. The armonica was the invention of which Franklin was most proud. Ironically, the ethereal sound of the armonica was used frequently by Franz Mesmer to relax his patients and provide the right atmosphere in which to work his magnetic magic.

The high-level commission formed by the French Academy of Sciences to investigate Mesmer contained not only Franklin but also the world-famous chemist Antoine Lavoisier. Their investigation contained various experiments, which included selecting a 12-year-old boy who was particularly susceptible to mesmerism, blindfolding him, and asking him to embrace various trees, one of which had been 'magnetized' by a practising mesmerist (not Mesmer, who had refused to cooperate with the commission). The boy went into convulsions after embracing a particular tree, but it was not the one which had been suitably anointed. After many related experiments, the commission concluded that the magnetic basis of mesmerism had no basis in fact, and that the reported effects were due simply to the power of suggestion. In their report, they summarized their findings as follows:

> The Commissioners having recognized that this animal-magnetism fluid cannot be perceived by any of our senses, that it had no action whatsoever, neither on themselves, nor on patients submitted to it; ... having finally demonstrated by decisive experiments that the imagination without magnetism produces convulsions, and that magnetism without imagination produces nothing

they had:

> unanimously concluded, on the question of the existence and
> utility of magnetism, that nothing proves the existence of
> animal-magnetism fluid; that this fluid with no existence is
> therefore without utility

Mesmer's technique showed only the power of the human
imagination and the important role that can be played by a
sympathetic and attentive practioner. It also highlighted the
nature of psychosomatic illnesses and the link between mind and
health. Mesmer's use of relaxation and inducing a trance-like state
(hence the word *mesmerize*) inadvertently laid foundations for the
subsequent use of hypnosis in therapy.

Reason and irrationality

In an age of modern evidence-based medicine, it might be thought
that magnetic healing would have completely disappeared. Look
in any modern bookshop and you will see that this is far from the
case. Many have a generously stocked section entitled 'Mind, Body,
Spirit' (though the classification 'Utter Nonsense' might be more
appropriate) in which one can find numerous titles discussing
magnetic healing or describing the supposed therapeutic power of
crystals. In one such volume, I found the assertion (unsupported
by any documented scientific evidence) that lodestones can be
used to 'channel energies' and 'reduce negativity', and that they
attack certain cancers and can combat diseases of the liver and the
blood. Such specious claims would not be out of place in a book
from the Middle Ages, but they can be found in books published in
the 21st century. Irrationality is alive and well and sold in a
bookshop near you.

The example of Franklin versus Mesmer can be seen as an early
triumph of reason against unreason and logic against
charlatinism. But such an emphasis of knowledge obtained and

demonstrated empirically substantially predated Franklin and in the field of magnetism no-one made a bigger contribution to the importance of rational deduction and experimental testing of hypotheses than William Gilbert. Writing at the end of the reign of Queen Elizabeth I, here is Gilbert describing his own philosophy of knowledge.

> Men are deplorably ignorant with respect to natural things and modern philosophers, as though dreaming in the darkness, must be aroused and taught the uses of things, the dealing with things; they must be made to quit the sort of learning that comes only from books, and that rests only on vain arguments from probability and upon conjectures.

It is to Gilbert's advances that we next turn.

Chapter 2
The Earth as a magnet

> But when the nature of the lodestone shall have been in the discourse following disclosed, and shall have been by our labours and experiments tested, then will the hidden and recondite but real causes of this great effect be brought forward, proven, shown, demonstrated; then, too will all darkness vanish; every smallest root of error, being plucked up, will be cast away and will be neglected; and the foundations of a grand magnetic science being laid will appear anew, so that high intellects may no more be deluded by vain opinions.
>
> William Gilbert, *De Magnete*, 1600

At the dawn of the 17th century, one book provided more clear thinking about magnetism than all the accumulated writings on the subject that preceded it. William Gilbert produced his masterpiece *De Magnete* in 1600, just three years before his own death. Born in Colchester in 1544, Gilbert had followed a degree in Cambridge with a career rising up the ranks of medical men, and when *De Magnete* was published he had just taken over as Queen Elizabeth's personal physician. He survived her and continued in post working for her successor, King James, no mean feat in those days of treacherous palace intrigue. He was, however, no stuffy courtier or staid traditionalist, slavishly following the wisdom of the ancients; Gilbert embraced the radical new theory of

Copernicanism and set his face against what he saw as the dead letter of Aristotelianism that had so dominated European thinking down the centuries. Moreover, in his writing, Gilbert didn't mince his words.

'Pretenders to science'

Right at the start of *De Magnete*, Gilbert set the tone for his whole work, stating that in former times 'philosophy, still rude and uncultured, was involved in the murkiness of errors and ignorances'. He then launched into a magnificent and spirited tirade against the errors of the past, and it is worth looking at some of these examples which he quotes to see the level of ignorance Gilbert was up against.

Gilbert stated that it had been asserted that a lodestone rubbed with garlic does not attract iron, nor does it work in the presence of diamond. Placed unawares under the head of a sleeping woman, a lodestone was claimed to 'drive her out of her bed if she be an adultress'. Allegedly, it could also make husbands agreeable to their wives, be used by thieves to open locks, or free women from witchcraft (why not men, one wonders?). Held in the hand, it 'cures pains of the feet and the cramps', or bestows eloquence. Some thought lodestones work only during the daytime (at night, the power is allegedly nulled); others held that the power of a weakened lodestone could be restored by the blood of a buck. Pickling a magnet 'with salt of the sucking-fish has the power of picking up a piece of gold from bottom of the deepest well'. Gilbert mercilessly mocked these examples, quoting each one with a detailed citation, and seems to have particularly enjoyed describing the work of one scholar, Lucas Gauricus, who thought that the lodestone belongs to the sign Virgo 'and with a veil of mathematical erudition does he cover many similar disgraceful stupidities'.

Gilbert did pay tribute to some of the ancients, the 'first parents of philosophy' such as Aristotle and Ptolemaeus, and felt that 'were they among the living' and if they could have seen his experiments they would have been firmly on his side. Even Thomas Aquinas, who was a believer in the effect of garlic on magnets, would have surely been persuaded, for 'with his godlike and perspicacious mind he would have developed many a point had he been acquainted with magnetic experiments'.

Even though lodestones used in the magnetic compass were well known to be useful, it was clear to Gilbert that no-one had anything worthwhile to say about how they actually worked. The magnetic compass needle points north because it is attracted to 'part of the heavens which overhangs the northern point', or alternatively to the pole star, a star in the tail of Ursa Major (some maintained that a large lodestone is located in the sky below the tail of Ursa Major), or possibly by being attracted to a range of magnetic mountains or a magnetic island at an unknown geographical location. Various legends of magnetic mountains abounded, so that it had been thought that ships needed to be constructed with wooden pegs so that as they sail by the magnetic cliffs 'there be no iron nails to draw out'. All of these ideas were thought by Gilbert to be 'world-wide astray from the truth and are blindly wandering'.

It was reported that lodestone had medicinal properties: it variously caused mental disturbance, melancholia, it preserved youthfulness, purged the bowels, or alternatively worked to 'stay the purging', corrected 'excessive humours of the bowels and putrescence of the same', and could be used to cure headaches or stab wounds. Gilbert caustically remarks: 'Thus do pretenders to science vainly and proposterously seek for remedies, ignorant of the true causes of things.' In this area at least, the medically qualified Gilbert did not think that iron was without healing power, though wisely he did not associate its efficacy with its magnetic powers. He acknowledged its use in treating some disorders of the liver and the spleen, noting that 'young women of pale, muddy, blotchy complexion are by it restored to soundness and comeliness'. Of course, iron tablets are today used to treat anaemia, and so Gilbert's instincts in this area were far from being way off the mark.

Gilbert's experiments

Perhaps because of the abundance of baloney concerning magnets that circulated in his day, Gilbert began his treatise by confessing to being rather wary of submitting his 'noble' and 'new' philosophy to 'the judgement of men who have taken oath to follow the opinions of others, to the most senseless corrupters of the arts, to lettered clowns, grammatists, sophists, spouters, and the wrong-headed rabble', particularly perhaps when he had described them in such insulting terms. However, he insisted that he was really addressing himself to 'true philosophers, ingenious minds, who not only in books but in things themselves look for knowledge'. Here is the key: his work was written for those who were no longer content with aping Aristotle and the ancients but to those who wish to look for truth 'in things themselves'. This then was the ace up his sleeve in his battle against the established order: 'Let whosoever would make the same experiments, handle the bodies carefully, skilfuly and deftly, not heedlessly and bunglingly.' In other words, if you don't believe Gilbert's reports, try the experiments yourself. But do so carefully. 'When an experiment fails, let him not in his ignorance condemn our discoveries', he said, since everything has 'been investigated again and again and repeated under our eyes'.

Experiment was Gilbert's weapon of choice, and so he devised and carried out numerous experiments on magnets. He found that a thin piece of pure iron, drawn out as a long wire, acts like a lodestone, magnetized along the length of the wire, an effect he saw in knitting needles and slender threads. He recognized that long pieces of iron which are heated, aligned north, and hammered by a blacksmith, and left to cool in that direction, will be magnetized along that direction. Gilbert found that some lodestones work better than others and that the shape is important, with oblong stones being more effective than spherical ones. He determined that pieces of iron can be magnetized using a lodestone, but rubbing a lodestone with other metals, wood, bone, or glass has no effect. Gilbert recognized that in the ancient world some silver coins had iron mixed in them by 'avaricious princes' and this could be the origin of reports that some lodestones could attract silver. But no such excuse could be made for the hypothesis of flesh-attracting lodestone or those that supposedly attract glass. He pointed out that lodestones can be used for magnetic separation, separating iron particles out from those of other metals.

Gilbert performed experiments also with electricity, playing with pieces of amber (fossilized tree resin, called *elektron* in Greek), rubbing them and observing the static electricity thereby produced using a pivoted needle of his own invention which he called a versorium – this was essentially the first electroscope. Gilbert's term 'electricus' (amber-like) was adopted half a century later into the English word *electricity*. Gilbert concluded that rubbing amber liberated an 'electric effluvium' which differs from air and was specific to the material being rubbed. This effluvium he held to be responsible for the electrical attraction. In fact, his effluvium was electrical charge and turned out to be present in all materials. Crucially, he argued that electricity and magnetism were distinct phenomena, though unfortunately his correct conclusion came from incomplete logic; he claimed that the electrical effect disappeared on general warming whereas magnetism does not. In fact, magnetism too is destroyed on heating, and this is something he should have known. He had recognized that a red-hot iron rod has no effect on a magnetized needle and similarly a magnetized piece of iron will lose its magnetism when placed in a hot fire and roasted until it glows red hot; the same effect was found to occur with iron filings.

In his most influential experiment, Gilbert procured a lodestone, and using his lathe he fashioned it into a sphere. He provocatively termed the resulting round magnet a 'terrella', literally a 'little Earth'. Passing a compass needle around the terrella, he found that the needle pointed in different directions as the compass needle moved around the sphere, and he realized that this behaviour mimicked the behaviour of a compass needle at different locations around the Earth (Figure 2). The logic was inescapable: a plausible mechanism for the origin of the magnetic effect on a compass needle was the magnetism of the Earth itself. Planet Earth behaves exactly like a giant terrella.

This picture goes some way to explaining the 'dip', or inclination, of the Earth's magnetic field, that is, why at different locations on

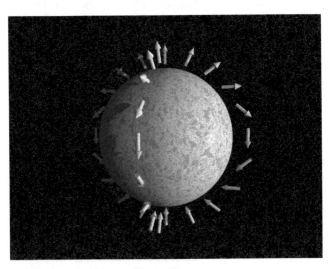

2. The magnetic field around a terrella

Earth the magnetic field lies at various angles to the horizontal. More difficult for Gilbert was the magnetic 'variation', that is, why the compass does not always point to true North but its precise direction varies slightly at different points on the globe. Gilbert came up with an elaborate explanation involving the magnetic effect of the mountains and continental land masses, but the data available to him were insufficient for him to realize this explanation is wrong. Magnetic variation is a subject we will return to in Chapter 9.

Petrus Peregrinus

Some of Gilbert's results had been anticipated by a 13th-century French scholar Pierre de Maricourt, more commonly known by his Latin name Petrus Peregrinus. His 'Letter on the Magnet', written to an acquaintance in his native Picardy, possibly when the author was a soldier in southern Italy, described an account of his own experiments on lodestones and included a detailed description of the freely pivoting compass needle. Peregrinus appears to have been motivated by his research to use magnets to construct a perpetual motion machine. Gilbert was passionately and rightly critical of any such venture. 'May the gods damn all such sham, pilfered, distorted words, which do but muddle the minds of students' is Gilbert's tart response.

Despite the utterly doomed nature of his quest, Peregrinus had performed some important experiments which were way ahead of their time. He showed that the poles of a magnet could attract or repel other poles, and he was in fact the first to describe a magnetic pole. His imagined mechanism for the compass was incorrect, though, as he guessed that the compass needle points to the celestial pole rather than, as Gilbert deduced, the terrestial poles of the planet.

Peregrinus also conducted an experiment that revealed that cutting a magnet in half produces two separate magnets, a north pole and a south pole appearing at the cut edges. Gilbert repeated this experiment, and to explain it he drew an analogy between this effect and the grafting together of two twigs of an easily sprouting tree such as willow; the two twigs could be grafted in either order but one has to maintain the sense of direction of growth in each case. Magnets similarly have a well-defined direction.

Why iron?

A common belief of the time was that the planets were associated with particular metals. They connected the Sun with gold, the Moon with silver, Venus with copper, and so on. Gilbert was not persuaded by these 'simpletons and raving astrologers' and asked 'what is common between Mars and iron, save that many other implements, swords and artillery are made of iron? What has copper to do with Venus? Or how does tin, or zinc, relate to Jupiter?' Gilbert gave a detailed description of how iron is extracted from iron ore (noting that a lodestone not only attracts iron but also iron ore) and reviewed its use in the making of tools, weapons, and utensils. In a lengthy hymn of praise to iron, Gilbert listed the metal's myriad uses:

> nails, hinges, bolts, saws, keys, bars, doors, folding-doors, spades, rods, pitchforks, heckles, hooks, fish-spears, pots, tripods, anvils, hammers, wedges, chains, manacles, fetters, hoes, mattocks, sickes ... forks, pans, ladles, spoons, roasting spits, knives, daggers, swords, axes ... strings of musical instruments, armchairs, portcullises, bows, catapults and those pests of humanity, bombs, muskets, cannon-balls ...

He went on to stress that every village had an iron forge and iron is 'far more abundant in the earth than the other metals'. This Gilbert saw as the knock-down argument against alchemy: 'it is a vain imagination of chemists to deem that nature's purpose is to change all metals to gold' or to 'change all stones into diamond'.

Gold may glitter and diamonds may sparkle, but to Gilbert it was obvious that iron is much more useful than gold or diamonds.

But what was special about iron? The Roman historian Plutarch thought that lodestone emitted some kind of heavy exhalation so that its magnetic influence was carried by ripples in the air, much as we now understand sound waves to be transmitted. Gilbert recognized that though various magnetic ores give off various noxious vapours when roasted, so do other ores, and hence magnetism cannot be due to them.

Gerolamo Cardano, a 16th-century mathematician and physician, thought that iron was special among the metals because of its excessive cold. Gilbert dismisses this as 'sorry trifling, no better than old wives' gossip'. Cornelius Gemma, a 16th-century astronomer (who incidentally provided the first illustration of the aurora and the human tapeworm, though probably not simultaneously), thought magnetism works by invisible rods; while Julius Scaliger, a staunch Aristotelian of the early 16th century, claimed that iron moves to a lodestone as to its mother's womb.

Both Cardano and also Alexander of Aphrodisias, an Aristotelian commentator from the late 2nd, early 3rd century, were so struck by the apparent life force animating magnets that they proposed that the lodestone actually feeds on iron. Giambattista della Porta did an experiment to test this idea, burying a lodestone in the ground together with a supply of tasty iron filings. Several months later, he dug them up, finding the lodestone a bit heavier and the filings correspondingly lighter, though the effect was very small. To say that Gilbert was very sceptical about this experiment is something of an understatement.

He did, however, have some sympathy with the life force argument and notes that most ancient philosophers declare the whole world to be endowed with a soul 'whereby the whole world blooms with most beautiful diversity'. However, he laments that Aristotle only ascribed such an animate nature to the heavens whereas the 'luckless' Earth is left 'imperfect, dead, inanimate, and subject to decay'. Gilbert's own view was more holistic, and he claimed that the 'Earth's magnetic force and the formate soul or animate form of the globes . . . exert an unending action, quick, definite, constant, directive, motive, imperant, through the whole mass of matter.'

The impact of *De Magnete*

Gilbert had produced a work of profound insight that had been based on experiment and observation and had used it to deduce not only facts about magnetism but also about the wider world. He had made the first step on the road to understanding the magnetism of the Earth. He correctly explained the tides as being due to the influence of the Moon, but incorrectly thought that the influence was mediated magnetically. Gilbert thus fell into the familiar trap that has ensnared many a genius and mere mortal alike, namely that when you get a good idea you tend to see it applying to absolutely everything.

Even though *De Magnete* was a technical treatise which dealt with rather abstract concepts, Gilbert's work became a runaway bestseller and succeeded in confirming magnetism as a fashionable topic of conversation in the early 17th century. However, magnetism was already all the rage in the popular culture of the day. Shakespeare's plays contain many references to magnetism. For example, in *A Midsummer Night's Dream* (written about five years before the publication of *De Magnete*), Helena, chased by Demetrius, exclaims:

> You draw me, you hard-hearted adamant;
> But yet you draw not iron, for my heart
> Is true as steel. Leave you your power to draw,
> And I shall have no power to follow you.

The imagery of iron drawn by the power of a magnet plays to the idea that the mystery of human emotional attraction is as unfathomable as the mystery of magnetic attraction. Ben Jonson's final comedy was entitled *The Magnetic Lady* and was first performed in 1632. It is a tale of the wealthy Lady Loadstone and her magnetically attractive niece Placentia Steel. The cast of characters includes a scholar, Mr Compass, the niece's nurse Mistress Keepe (magnets were often sold with a keeper that was placed over the pole pieces when the magnet was not in use in order to preserve the life of the magnet), a soldier Captain Ironside, and a tailor Mr Needle. This was probably taking the whole thing rather too far!

But Gilbert's lasting legacy was more than just establishing that the Earth is a magnet. Most importantly, Gilbert made experiments fashionable. The way to make progress in this subject was not to pontificate, or even to write plays, but to devise and carry out experiments. The following chapter is all about people who did precisely that.

Chapter 3
Electrical current and the path to power

Gilbert had understood the principle of the magnetic compass, but had not come up with a complete understanding of what magnets actually are. He had recognized that magnetism and electricity were different phenomena, pieces of lodestone and pieces of amber behaving quite distinctly, but the connection between these two effects eluded him. This is hardly a poor reflection on Gilbert as it would take a few hundred years for this connection to be appreciated by anyone else.

In the intervening period, the celebrated French philosopher René Descartes attempted to understand magnetism. Descartes conjured up an ingenious model of *spiral effluvia* which are, as he explained in a letter in 1643 to the Dutch physicist Christiaan Huygens, 'a very subtle and imperceptible kind of matter which emerges continuously from the Earth, not from the pole, but from every part of the northern hemisphere, and then passes to the south, where it proceeds to enter every part of the southern hemisphere'. Descartes' spiral effluvia find their travel through space tiresome and happily take a detour through any pieces of lodestone on their way. Descartes' musings were entertaining, and were accompanied by an attractive diagram (Figure 3) which gives some indication of how he pictured things working, but his ideas were entirely divorced from experiment. However, others did

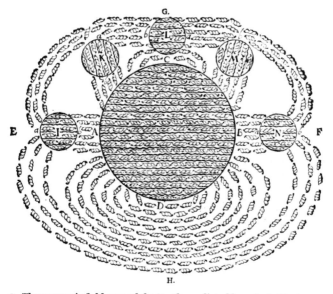

3. The magnetic field around the Earth, mediated by spiral effluvia, according to Descartes

follow Gilbert's tradition of experimentally based investigations of magnetism. This chapter will describe the beginnings of the modern understanding of the connection between electricity and magnetism, and this came from a series of decisive experiments, from Galvani's work on frogs through to the discoveries of Ampère, Faraday, and Tesla.

Animal electricity

In Bologna, 171 years after the publication of *De Magnete*, Luigi Galvani discovered that the leg muscles of dead frogs twitch when subjected to electrical sparks or impulses. This was a clue that animals 'worked' like electrical machines, and very soon he and his nephew and assistant, Giovanni Aldini, were able to extend the experiment to the muscles in a variety of other dead animals.

Before long, the demonstration became something of a macabre party piece: sets of frogs' legs wired together would flinch in synchrony, and the severed heads of chickens, sheep, and oxen could also be made to twitch under electrical control. Aldini toured Europe with this travelling show, amazing and delighting packed houses. When Aldini's tour reached London in 1803, the body of an executed murderer was taken from the gallows at Newgate Prison and sent to the Royal College of Surgeons where Aldini passed electricity through the corpse's face. It was reported that 'the jaw began to quiver, the adjoining muscles were horribly contorted, and the left eye actually opened'. The conducting leads attached to other parts of the body caused different spasms, the clenching of a fist, the kicking of the legs, and arching of the back. Electricity now appeared to be the force which breathed life into all types of organisms, an effect vividly captured in Mary Shelley's 1818 novel *Frankenstein*, in which, at least in later film versions, the monster is created by knitting together body parts and reanimating it using electricity.

We now understand that nerve impulses are indeed electrical in nature and that an external electric potential can be used to produce a measurable response. This is an effect that is put to good use in a heart pacemaker to precisely regulate the beating of a malfunctioning heart. Electrical activity in the brain can also be altered in a treatment known as electroconvulsive therapy, sometimes used to treat major depression, though the efficacy is harder to predict. The physicist Alessandro Volta referred to Galvani's animal electricity as galvanism and the name has stuck; we still talk about an audience hearing a rousing speech being 'galvanized into action'.

Galvani was also able to demonstrate that lightning in thunderstorms was electrical by constructing a lightning detector made out of frogs' legs. One end of each leg was connected to a vertical metal rod which was exposed to the elements, the other end being connected to a wire which ran down a well (and was thus grounded). When lightning flashed, the frogs' legs twitched (and the thunder came noticeably later), thus demonstrating that lightning was electrical.

The battery

Alessandro Volta continued these sorts of experiments but began to use live frogs. In one landmark experiment, he found that he could dispense with an external generator of electricity (a spark generator or the Leyden jar we will encounter shortly) if the circuit consisted only of the frog and two dissimilar metals. This seemed to be proof that the poor frog was itself the source of electrical power, a vindication of the notion of animal electricity. However, subsequent investigation showed that it was the two dissimilar metals (connected by a wet interface) that did the trick. Volta had invented what soon became known as the voltaic cell, now called simply a battery (whose strength is measured in volts). Frogs were no longer needed in electrical circuitry (indeed there shouldn't be one inside your iPhone).

The battery dispensed with the Leyden jar as the simplest system for producing electricity in a laboratory. Invented in the 1740s in Leiden (then spelled Leyden), the Leyden jar consisted of a large glass vessel with its inner and outer surface coated in metal foil and partially filled with water. A metal rod located along the axis of the jar emerges through the mouth of the jar and is connected to the inner foil at the bottom. The jar could store static electricity and, as mentioned earlier, Benjamin Franklin was the one to demonstrate this was due to charge residing on the glass surfaces (the Leyden jar was essentially a large capacitor). Volta's batteries were much more convenient to work with and permitted the new science of electricity to be investigated much more easily.

A battery has a positive and a negative terminal, and this rather resembles a magnet which has a north and a south pole. (This was in itself seen to be a vague hint that electricity and magnetism might be connected.) At the end of the 18th century, the French physicist Charles-Augustin de Coulomb had shown that both electrical charges and magnetic poles produced an influence which varied in inverse proportion to the square of the distance from them (the inverse-square law), and this was further evidence for some underlying connection between the two phenomena. Coulomb's explanation for the origin of magnetism rested on the supposed existence of two types of magnetic fluid (boreal and austral, i.e. north and south) which were imprisoned in magnets, for some reason that was never made clear.

Hans Christian Oersted: a current produces a magnetic field

In April 1820, Hans Christian Oersted, a professor of physics at the University of Copenhagen, made what turned out to be the crucial observation. He noticed that a compass needle placed close to a wire shows a sudden deflection when an electric current through the wire was switched on and off. The effect was not large, and Oersted only observed a faint twitch in the compass needle. He presented the effect in a lecture, later recording that 'as the effect was very feeble... the experiment made no strong impression on the audience'. But it was enough to make an impression on Oersted. The needle clearly and repeatedly twitched back and forth as the current was switched on and off. Subsequent experimentation showed that the magnetic field produced by the current flows around the wire as shown in Figure 4. It took Oersted a while to demonstrate it because he had originally imagined that any possible effect would be produced parallel to the wire and looked for it in that geometry. It was extremely surprising that the effect along the wire was entirely absent, but showed itself in directions perpendicular to the flow of the current. Furthermore, the magnetic field lines did not radiate away from the wire, but flowed around it in circulating loops. This was an extraordinary discovery and, while a connection between

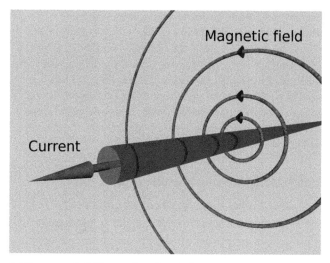

Magnetic field

Current

4. The magnetic field around an electrical current

electricity and magnetism had been broadly expected, this
circulating loop effect was a surprise.

The circulating nature of the magnetic field around the current
was particularly interesting because it showed a definite
handedness: magnetic field circulates in one sense only around
the current. Oersted described this effect as a 'dextrorsum spiral', a
botanic term that normally refers to the helicity of climbing plants
which also wind round with a particular handedness. Oersted's
colourful phrase has sadly not survived, but students now learn
the sense of the winding of the magnetic field using Fleming's rule,
a handy mnemonic dreamed up by Sir John Ambrose Fleming
(who also invented the thermionic valve, see Chapter 8).

The significance of Oersted's discovery of an electrically generated
magnetic field was considerable. Hitherto, magnetic fields had
only been possible to produce using a lodestone, which is of course
a naturally occurring magnetic mineral hewn from the rock. Now

magnetic fields could be manufactured artificially in the laboratory. All you needed was a battery and a wire. Inserting a switch into the circuit meant that you could turn the magnetic field on and off whenever you wanted. No piece of lodestone came fitted with a switch: lodestones are 'on' all the time. In fact, you could go further with this argument. Using a variable resistor (like a dimmer control) in your circuit meant that you could turn the strength of the magnetic field up and down, another feature not available with the lodestone.

André-Marie Ampère and electrodynamics

André-Marie Ampère heard about Oersted's discovery in September 1820 and began to work on trying to understand it, a demanding task since Oersted's account was rather sketchy and contained no diagrams (though Oersted's achievement was considerable as the effect he measured was comparable with the magnetic field from the Earth, the effect of which he had needed to subtract in his head). Ampère taught at the École Polytechnique in Paris. He had an excellent background in mathematical physics, but was not above getting his hands dirty and performing experiments himself. He reasoned that since magnets can attract or repel each other, and Oersted had shown that a current-carrying wire produces a magnetic field, maybe current-carrying wires might attract or repel each other. To test this hypothesis, Ampère designed a sensitive balance in order to measure any force between current-carrying wires and was able to demonstrate that a very weak force indeed existed. The force was attractive if the two wires carried current in the same direction; repulsive if the current flowed in the opposite direction. He formulated a law which described this new effect, stating that the force on two current-carrying lengths of wire was proportional to their lengths and proportional to the flow of the current in each wire. Ampère also speculated on the origin of magnetism in matter, and since Oersted had observed that an electrical current produces a magnetic field, it was but a short step to deduce that a magnet might itself contain elementary, microscopic electric currents. Ampère therefore pictured a magnet as being dynamically alive, brimful of continually flowing microscopic currents that caused the magnetic field surrounding it. Ampère called his theory *electrodynamics*.

By 1826, Ampère was ready to announce a mathematical derivation of his electrodynamic force law, and did so in his *Memoir on the Mathematical Theory of Electrodynamic Phenomena Deduced Solely from Experiments*. In modern notation, Ampère's law, which derives from this work, is expressed as the magnetic field added together around a loop encircling a wire being proportional to the current in that wire. It therefore provides a precise mathematical description of Oersted's experiment.

One disadvantage with the Oersted experiment was that the magnetic field from a current-carrying wire is actually rather weak and becomes weaker with increasing distance from the wire. Ampère found an ingenious way to magnify the effect, namely to wind the wire into a coil, essentially allowing a single wire to produce many copies of its magnetic field and add them up to make something bigger. The magnetic field inside a coil is rather uniform in strength and can be much more intense, being proportional to the number of turns of wire wound on the coil per unit length. Such a device is called a solenoid, derived from the Greek meaning 'in the form of a pipe'. Another way of magnifying the magnetic field was to wind the coil around a horseshoe-shaped piece of iron. This allows the coil to magnetize the iron and the field between the poles of the magnet can be extemely intense. One application of these ideas was the development of the electromagnet. By coiling a copper wire around an iron horseshoe magnet, William Sturgeon in 1824 produced a machine in which the magnetic field between the poles of the horseshoe could be controlled by applying a current. From the late 1820s onwards, larger and larger electromagnets were constructed, with Joseph

Henry in the US popularizing their use. He took great delight in repeatedly breaking the record for the largest weight lifted by his increasingly enormous and powerful electromagnets. Henry's particular innovation was to insulate the wires coiling around the iron cores; this allowed more turns of coiled wire to be wound around the iron cores, further increasing the field strength.

Michael Faraday and lines of force

Some of the most important experiments ever performed in magnetism are due to Michael Faraday. He got his big break when, as a 20-year-old apprentice bookbinder from a modest background, he presented the eminent chemist Sir Humphry Davy with a bound version of the notes he had taken at some of Davy's public lectures, and thereby secured a job as Sir Humphry's scientific assistant at the Royal Institution in London. Faraday eventually succeeded Davy as Director of the Royal Institution and spent his life devoted to scientific investigation. While Davy was still Director, news reached London of Oersted and Ampère's work, and it was obvious that here was an exciting opportunity. Oersted had found that an electric current produces a magnetic field, and Ampère had shown that two current-carrying wires produce a force on each other, so using magnetism there was a potential route from electricity to mechanical work. The Royal Institution had the benefit of larger and more powerful batteries than available to Ampère and so Faraday was able to repeat Ampère's experiments easily.

Faraday then performed an experiment which showed that a bar magnet produced a force on a current-carrying wire and, intriguingly, the force was found to be perpendicular to both the direction of the magnetic field (as measured by a compass needle) and the direction of the current. This property gave Faraday an idea for a new type of machine. He reasoned that a current-carrying wire might be able to keep a magnet revolving in a continuous circular motion if he could arrange things

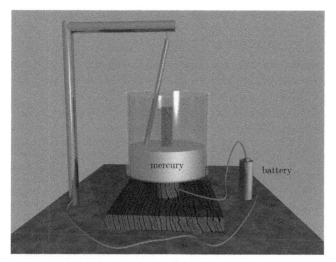

5. Faraday's electric motor

appropriately. He then set to work, fixing a bar magnet vertically in the bottom of a cylindrical beaker and filling the beaker with mercury (see Figure 5). A wire dangled from a pivot above the beaker and touched the surface of the mercury, but hung at an angle to the vertical. A current flowing through the wire then caused it to rotate around the magnet. This was because the magnet produced a force on the wire at right angles to its length. Faraday had produced the first electric motor. He rushed into print to announce his discovery, and in his haste he neglected to acknowledge discussions he had had with his boss. Together with William Wollaston, Sir Humphry Davy had been trying unsuccessfully to build an electric motor, and Faraday had discussed the problem with both men. Outsmarting his superiors was one thing, but failing to at least acknowledge their discussions was bad politics to say the least, and Faraday's relationship with Davy was severely strained.

Notwithstanding his difficulties with his colleagues, Faraday had produced what is now known as a homopolar motor, so named because the current always flows in the same direction. His device was little more than a toy, but it demonstrated the principle of conversion of electrical energy into mechanical energy. Later designs of motor soon incorporated what is known as a commutator, a kind of rotary switch that periodically reverses the direction of the current flow as the motor rotates.

The next breakthrough was another discovery made by Faraday, that of electromagnetic induction, which he made in 1831. The reasoning that led to this idea was that if, as Oersted had demonstrated, an electrical current might produce a magnetic field then, perhaps, the opposite might be true: a magnetic field might be used to *induce* an electric current. Faraday did just that and found that a current flowing in a coil wound round one side of an iron ring could induce a current in a second coil wound the other side. However, the induced electric current was transient, and was completely absent when the current flowed steadily in the first coil. It only appeared in the second coil for a short interval after the current in the first coil was switched on. Rather more surprising for Faraday was the observation that another transient current was induced in the second coil when the current in the first coil was disconnected. (An effect while turning on a current was perhaps conceivable, but seeing an effect when the current was rapidly dying in the circuit was astonishing.) What Faraday appeared to be seeing in his experiment was that a current was being induced in the second coil only when the magnetic field in the iron ring was *changing*.

To understand this idea, it is most helpful not to consider Faraday's original example but something more idealized. Imagine a metal wire moving vertically downwards through a stationary magnetic field. The wire contains electrical charges, but the positive and negative charges balance out. Because the wire moves vertically downwards, the charges do so too, and the magnetic field exerts a force on them, but the direction of the force depends on the sign of the charge (to the right for positive charges, to the left for negative charges). This means that a current is induced in the wire. (In a real metal, only the negative charges in a wire are mobile, but that does not change the argument.) It seems as if the motion of the wire has induced a voltage in the wire to drive the current. We now have a new way of generating a voltage; in addition to the battery, we can also move a wire through a magnetic field. However, the effect also works if we keep the wire stationary and move the magnet which causes the magnetic field!

Now imagine a circle of wire, broken at one point and connected to a voltmeter. A steady magnetic field is applied perpendicular to the circle. No voltage will be read by the voltmeter since there is nothing else in the circuit. So far, no surprise. Now allow the magnetic field to vary with time. The voltmeter will now flicker into life. As the magnetic field increases, a voltage has been induced in the wire. The voltage will depend on the rate at which the magnetic field is changed and is said to be *induced* by the changing magnetic field.

You can change the magnetic field passing through a loop in more than one way. For example, you can keep the magnetic field constant and allow the loop of wire to move through space. To increase the effect, you can take a coil of wire rather than a single loop, and the voltage induced increases in proportion to the number of turns on the coil. Faraday realized that this effect could be used to transform mechanical energy (turning a magnet or a coil of wire) into electrical energy (the induced voltage) and hence produce an entirely new way to generate electrical power. Though Faraday had made his discovery of *electromagnetic induction* in 1831, Joseph Henry, later to be the first Curator of the Smithsonian Museum in Washington, DC, performed similar work in Albany, New York, at around the same time.

Faraday first demonstrated electromagnetic induction using a rotating metal disk in a fixed magnetic field, but this was not the only way to do it. The hand-cranked rotating machine of the French instrument-maker Hippolyte Pixii was constucted in 1832 and worked on the same principle. Before long, many such dynamos had been constructed (in fact, the design of the dynamo had been anticipated by an invention of the Hungarian engineer Ányos Jedlik, who in 1827 had designed and built a machine with a commutator and other parts of a modern motor).

Faraday was not only a prolific discoverer of scientific effects but also a great originator of scientific terms. We have him to thank for the following: cathode, anode, electrode, cation, anion, ion, and electrolyte. He created these new words with great care and frequently under the kindly but firm tutelage of William Whewell, the Cambridge philosopher, scientist, and theologian who politely vetoed suggestions he thought were clumsy and frequently steered Faraday into adopting his own instead. Whewell is himself responsible for giving us the words 'scientist' and 'physicist' (the latter term Faraday intensely disliked because it is difficult to pronounce).

Faraday made all his remarkable scientific advances despite a complete lack of mathematical training and ability which caused him considerable embarrassment. Writing to Ampère, he half-apologized, regretting that his 'deficiency in mathematical theory' made him 'dull in comprehending these subjects'. He confessed: 'I am unfortunate in a want of mathematical knowledge and the power of entering with facility into abstract reasoning. I am obliged to feel my way by facts closely placed together, so that it often happens I am left behind in the progress of a branch of science'. Nevertheless, his more intuitive approach often brought dividends, leading him to comment: 'It is quite comfortable to find that experiment needs not quail before mathematics but is quite competent to rival it in discovery.'

AC/DC: the battle of the currents

In the 19th century, the principles were established on which the modern electromagnetic world could be built. The electrical turbine is the industrialized embodiment of Faraday's idea of producing electricity by rotating magnets. The turbine can be driven by the wind or by falling water in hydroelectric power stations; it can be powered by steam which is itself produced by boiling water using the heat produced from nuclear fission or burning coal or gas. Whatever the method, rotating magnets inducing currents feed the appetite of the world's cities for electricity, lighting our streets, powering our televisions and computers, and providing us with an abundant source of energy. Faraday's discovery has transformed the planet and rotating magnets are the engine of the modern world.

By the closing decades of the 19th century, the understanding of Faraday's law of induction had led to the widespread production of electrical power and the possibility that this new energy source could be piped directly into homes. One of the scientists who helped that to become a reality was Nikola Tesla, an obsessive and driven Serbian genius whose extraordinary creativity in designing electrical contraptions was matched only by his bizarre eccentricities. Tesla was obsessed by performing tasks in groups of threes, was fanatical about hygiene, yet in later life befriended pigeons in Central Park, New York, taking some of them back to his apartment. He was one of the first people to appreciate that practical electrical power distribution was best achieved with alternating current (AC), rather than direct current (DC), that is, with a voltage that oscillated back and forth rather than one that was fixed at a steady value. Very early in his career, Tesla conceived how to make an alternating current motor or generator using a rotating magnetic field produced by three coils oriented at 120 degrees from each other. If these coils were each fed with alternating current 120 degrees out of phase with each other, the resulting magnetic field rotates and can be used to generate torque in a machine. This is the principle behind the alternator, induction motor, and other alternating-current generators and underpins much of modern power technology. On moving to the United States in 1884, these ideas had yet to be realized and Tesla got his first job working for the American inventor Thomas Alva Edison, who recognized Tesla's talents but failed to promote him. Tesla subsequently set up his own company in direct opposition to Edison's, but Tesla was cheated by his business partners and his company flopped. Eventually, he forged an uneasy alliance with George Westinghouse, another Edison rival, and this new alliance led to the commercial production of his alternating-current motors.

Thomas Alva Edison is of course rightly celebrated as a brilliant inventor, but he had an unfortunate knack of not understanding the commercial significance of what he'd invented. Recording his voice (famously intoning 'Mary had a little lamb') onto a rotating wax cylinder heralded the advent of home music systems, but he thought his 'phonograph' had an application only as an office dictating machine. Edison's work on developing practical electric

6. Thomas Edison and Nikola Tesla

light should have made him a multi-millionaire, but here he became wedded to installing such systems using direct current. Edison was no theoretical physicist and the mathematics necessary to understand alternating current were beyond him; Tesla, in contrast, was a brilliant and intuitive mathematician who could immediately see the advantages of alternating current. Edison's competitors, including the company founded by Westinghouse, could provide home power systems which generated alternating current and were far cheaper.

Direct-current systems involve a large drop in voltage between the source of power generation and the place where it is needed, so the power plant has to be located fairly close to your home and the wires need to be of large diameter to reduce the electrical resistance, pushing up the cost. Alternating-current systems, on the other hand, can transmit power over large distances at high voltages (but low current, reducing the diameter of the wire) and then, close to where the power is needed, it can be transformed down to lower voltage.

The need for domestic electrical lighting was the major reason why it was important to find methods to deliver power directly to peoples' homes. Though today we rely on electricity to power numerous appliances, from computers to refrigerators, mobile phone chargers to televisions, these devices not only didn't exist but weren't even imagined at that time. Cooking, making toast, and boiling water could all be performed easily by other methods. But the need for electric light, the very technology that Edison had pioneered, provided the principal motivation for getting power piped across cities.

Edison distrusted alternating current because it necessitated higher voltages, and he insisted this made it much more dangerous. To demonstrate the danger, he ordered some of his technicians to use alternating current to electrocute various animals, mainly stray cats and dogs. When Edison was consulted by the State of New York on the best way to use electricity to execute prisoners, despite his personal opposition to capital punishment, Edison immediately recommended alternating current. Edison's endorsement was cynically motivated by a desire to discredit his enemies by fixing the danger of alternating current firmly in the minds of the American public. Edison could be ruthless, and it was once said of him that he had 'a vacuum where his conscience ought to be'. In the campaign to disparage their rivals, one of Edison's employees went even further than his boss by suggesting that they should introduce new terminology for the process of electrical execution, saying that the authorities would 'Westinghouse' the condemned person. Electrocution was, of course, the name that stuck, but alternating current was indeed used in the first electrical execution in 1890, though two successive attempts were needed to finish off the unfortunate prisoner.

However, despite these deeply distasteful tactics, Edison lost the battle of the currents. Westinghouse built working power plants using alternating current, the one at Niagara Falls in 1896 being a conspicuous and widely reported success, and Tesla's technology became dominant. Edison wasn't the only loser in the battle of the currents: Albert Einstein's father ran a company in Munich that manufactured equipment that ran on direct current, and following the rise of AC the company went out of business when Albert was 15 years old, forcing the family to move to Italy.

Modern society is built on the widespread availability of cheap electrical power, and almost all of it comes from magnets whirling around in turbines, producing electric current by the laws discovered by Oersted, Ampère, and Faraday. But what is the underlying connection between electricity and magnetism? We will consider this in the next chapter.

Chapter 4
Unification

Electricity and magnetism seem to be two separate phenomena, yet Oersted, Ampère, and Faraday's work all showed that there was a connection between them. The development of the motor and the generator showed that this electromagnetic connection could be harnessed to provide useful technology. But so far, we have simply stated some discovered laws and not discussed the fundamental origin of the connection between electricity and magnetism. That insight came through the work of James Clerk Maxwell whose unification of electricity and magnetism was one of the most beautiful, subtle, and imaginative developments in theoretical physics. Not only did it achieve the unification, it did so by explaining what light is.

James Clerk Maxwell

Born in Edinburgh in 1831, James Clerk Maxwell was brought up in the Scottish countryside at Glenair. He was educated at home until, at the age of 10, he was sent to the Edinburgh Academy where his unusual homemade clothes and distracted air earned him the nickname 'Dafty'. But Maxwell was far from daft, as demonstrated when he wrote his first scientific paper aged only 14. Maxwell went to Peterhouse, Cambridge, in 1850, and then moved to Trinity College, where he gained a fellowship in 1854. There he worked on the perception of colour, and also put Michael Faraday's ideas of lines of electrical force onto a sound mathematical basis. In 1856, he took up a chair in Natural Philosophy in Aberdeen where he worked on a theory of the rings of Saturn (confirmed by the *Voyager* spacecraft visits of the 1980s) and, in 1858, married the college principal's daughter, Katherine Mary Dewar.

In 1859, he was inspired by a paper of the German physicist Rudolf Clausius on diffusion in gases to conceive of his theory of speed distributions in gases, still used to this day. These triumphs were not enough to preserve him from the consequences of the 1860 merging of Aberdeen's two universities when, quite incredibly, the authorities decided that it was Maxwell out of the two Professors of Natural Philosophy who should be made redundant. He unaccountably failed to obtain a chair at Edinburgh but instead moved to King's College London. There, he produced the world's first colour photograph, and crucially came up with his theory of electromagnetism which described all the phenomena so far discovered, and furthermore proposed that light was an electromagnetic wave.

With a keen mathematical ability, Maxwell was well placed to make progress on this important topic. Although highly adept with equations, Maxwell made his mathematical manipulations subservient to physical insight. In this he took his cue from the experiments of Faraday rather than the formalism of the mathematicians. Writing in his *Treatise on Electricity and Magnetism*, Maxwell insisted that 'many of the mathematical discoveries of Laplace, Poisson, Green and Gauss find their proper place in this treatise, and their appropriate expression in terms of conceptions mainly derived from Faraday'. In other words, Faraday's entirely non-mathematical insights gleaned from experiment were the only rational starting point. Maxwell noted that he was 'aware that there was supposed to be a difference between Faraday's way of conceiving phenomena and that of the mathematicians, so that neither he nor they were satisfied with each other's language. I had also the conviction that this discrepancy did not arise from either party being wrong.' This was an understanding Maxwell obtained from consultation with the physicist William Thomson. Faraday's experiments on the patterns produced by iron filings sprinkled around a magnet gave him an actual map of the 'lines of force' as they curved around the magnet. In 1849, the young William Thomson had introduced the term 'field of force', or simply *field*, to describe this array of lines of force. Faraday rejected the abstract notions of 'action at a distance' and embraced this magnetic effect whose influence fills and pervades all space. He knew it was there because he could see its effects, and was considerably encouraged in his instinctive idea when Thomson put some mathematical flesh on these conceptual bones. Both Maxwell and Thomson were receptive to Faraday's insights because of their fascination with another branch of science, the movement of fluids.

Alfred, Lord Tennyson memorably described the flow of water in his poem 'The Brook' in which the stream itself describes what it feels like to bubble and babble on its path past 'field and fallow' to the 'brimming river'.

> I slip, I slide, I gloom, I glance,
> Among my skimming swallows;
> I make the netted sunbeam dance
> Against my sandy shallows.

Both Thomson and Maxwell had made contributions to the theory of fluid dynamics and thus had the mathematics necessary for describing the flow of fluid using a field of vectors, little imaginary arrows filling all of space, referring to the velocity of the fluid at each point. The motion of fluid, the slipping and sliding and glooming and glancing, could then be described by equations, providing a model to understand how the fluid flows along streamlines and sometimes swerves around vortices of swirling liquid. Thomson and Maxwell were thus particularly receptive to the idea that electric and magnetic fields filled space and could be modelled as a vector field, analogous to the one used to describe the velocity in a fluid. The motivation came from Faraday and his experiments. Maxwell devoured Faraday's *Experimental Researches in Electricity*, describing them appreciatively as:

> a strictly contemporary historical account of some of the greatest electrical discoveries and investigations, carried on in an order and succession which could hardly have been improved if the results had been known from the first, and expressed in the language of a man who devoted much of his attention to the methods of accurately describing scientific operations and their results.

Maxwell continued:

> As I proceeded with the study of Faraday, I perceived that his method of conceiving the phenomena was also a mathematical one, though not exhibited in the conventional form of mathematical symbols. I also found that these methods were capable of being expressed in the ordinary mathematical forms, and thus compared with those of the professed mathematicians.
>
> For instance, Faraday, in his mind's eye, saw lines of force traversing all space where the mathematicians saw centres of force attracting at a distance: Faraday saw a medium where they saw nothing but distance: Faraday sought the seat of the phenomena in real actions going on in the medium, they were satisfied that they had found it in a power of action at a distance impressed on the electric fluids.
>
> When I had translated what I considered to be Faraday's ideas into a mathematical form, I found that in general the results of the two methods coincided, so that the same phenomena were accounted for, and the same laws of action deduced by both methods, but that Faraday's methods resembled those in which we begin with the whole and arrive at the parts by analysis, while the ordinary mathematical methods were founded on the principle of beginning with the parts and building up the whole by synthesis.

I also found that several of the most fertile methods of research discovered by the mathematicians could be expressed much better in terms of ideas derived from Faraday than in their original form.

Maxwell's equations

Maxwell's theory of electromagnetism can be summarized in four beautiful equations. This is not a book containing equations, but Maxwell's equations are so important we will at least describe them. They describe the behaviour of electric and magnetic fields and how they relate to charges and currents.

The first of Maxwell's equations says that every line of electric field originates on a positive charge and ends up on a negative charge. Positive charges look like little factories of electric field, with electric field lines diverging away from them, while negative charges are consumers of electric field, with electric field lines diverging into them (see Figure 7(a) and (b)). Stated in something close to a mathematical expression (if you want the 'real thing', please consult the Mathematical Appendix), we could write the first equation:

(1) Divergence of electric field = amount of electrical charge

This means that if the charge is positive in some region, the divergence of electric field is positive. If the charge is negative, the

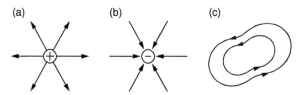

(a)   (b)   (c)

7. Electric field lines (a) diverge away from positive electrical charges and (b) diverge into negative electrical charges. (c) Magnetic field lines can only exist in loops; they never start or stop anywhere

divergence of electric field is negative. (The 'divergence' is a term that has a precise technical definition from which I will spare you, though if interested consult the Further reading.) By writing this equation, Maxwell was building on the work of the German mathematician Carl Friedrich Gauss, who provided a mathematical formulation of this very problem, and Maxwell's first equation subsumes what is known as 'Gauss' theorem'. To reiterate, Maxwell's first equation states that wherever there is a charge, then you will measure electric field lines diverging away from or into it; if there is a region of space in which there are no charges, then there will be no divergence of electric field.

Maxwell's second equation states simply that in magnetism there is no analogue of electric charge. Magnetic charges (sometimes known as magnetic monopoles) do not exist (a point we will return to in Chapter 10, but for the moment let us take it as a given), and by analogy with electric fields this would imply that there is no point away from which or into which magnetic fields diverge. Thus magnetic field lines cannot stop or start anywhere but must endlessly circulate in loops (see Figure 7(c)). This is exactly what had been observed in the experiments of Michael Faraday and others. The equivalent expression is then written as follows:

(2) Divergence of magnetic field = 0

But what determines how these fields circulate round and round? This is the subject of the second pair of equations. Let us start with electric fields. We have said that they simply radiate away from positive charges or converge towards negative charges, so they don't circulate. However, they can be made to circulate if there is a changing magnetic field present, as shown by Faraday's law of electromagnetic induction. Crudely, we can write this as a verbal equation:

(3) Circulation of electric field = changing magnetic field

For a magnetic field, the circulation is produced by an electrical current, a flow of charge, as shown experimentally by Ampère. Thus we write the fourth equation as:

(4) Circulation of magnetic field = current

At this point, Maxwell's four equations have summarized nothing more than was already known. However, Maxwell made the following brilliant leap in the dark. A changing magnetic field was known to produce an electric field (that is the induced voltage of Faraday). What if a changing electric field produced a magnetic field? Maxwell realized that if this were the case, then an extra term should be inserted in equation 4, changing it as follows:

(4') Circulation of magnetic field = current + changing electric field

This turned out to be the final piece of the jigsaw. It was already known that electricity and magnetism were connected, but it was not known how. When he surveyed this collection of equations, Maxwell realized that putting these equations together in one package allows you to see how these connections operate together. Imagine that you somehow get a current to oscillate up and down a wire. That will produce an oscillating magnetic field (by equation 4) and the oscillating magnetic field will induce an electric field (by equation 3). That electric field, existing in the space around the wire, will vary with time and hence produce a magnetic field (by equation 4), which will vary with time and produce an electric field (by equation 3), and so on. Maxwell realized that these changes in electric field will produce changes in magnetic field, and vice versa, and a self-sustaining wave of varying electric and magnetic fields will propagate off into space. Maxwell had predicted the existence of an electromagnetic wave. (In fact, the oscillating current in the wire that caused the whole thing in the first place is nothing more than a radio transmitter.)

But what speed would this wave travel at? To answer this, all one has to do is to solve the wave equation that results from welding Maxwell's equations together, a relatively simple procedure that physics undergraduates are regularly asked to repeat. However, it was not so simple for Maxwell himself to do. The main problem he faced was that the units used at that time treated electrical quantities and magnetic quantities entirely differently and the units were completely incompatible. (Imagine trying to work out how long it would take to drive to a distant town if your car's speedometer was calibrated in furlongs per fortnight.) Maxwell had hit upon his idea of an electromagnetic wave while he was spending his summer on his estate in Scotland, but his tables of units were back in his office in Cambridge. He had an agonizing wait for the end of his vacation before he could return and perform an accurate calculation. When he did so, he was delighted to find that the speed of the electromagnetic wave was precisely equal to what the French physicist Hippolyte Fizeau had measured for the speed of light.

The speed of light

Fizeau's 1849 apparatus was improved by Léon Foucault the following year, and a simplified version of the Foucault adaptation will be described. Light was propagated towards a rotating mirror from which it reflected to a second mirror some 35 kilometres away. From there, it was reflected back to the rotating mirror, but in the time it takes to perform the 70-kilometre round trip the rotating mirror will have rotated by a small angle, and so the light beam emerges along a different path from that at which it entered. For example, if the rotating mirror is spun at 10 revolutions per second, the small angle will be almost a degree, which is easily measurable.

Fizeau and Foucault were not the first to measure the speed of light, a record that probably belongs to Danish astronomer Ole Rømer who in 1676 noticed that the orbital period of Io, one of

Jupiter's moons, seemed to depend very slightly on time and in particular on whether the Earth was moving towards or away from Jupiter. The timing effect, Rømer reasoned, could be explained if light from Jupiter and its moons was taking longer to reach the Earth when they were further away than when they were close. The effect allowed him to make an estimate of the speed of light. The Fizeau–Foucault measurement was much more accurate, and also much more direct, than Rømer's, so Maxwell knew he had a reliable experimental value to check his theory.

You can begin to appreciate the astonishing nature of what Maxwell had deduced by writing his formula for the speed of light, c, in modern units. His theory predicted that the speed of light should be connected to the permittivity of free space (a quantity you can measure by studying an electrical capacitor and given the symbol ϵ_0) and the permeability of free space (measured using a magnetic solenoid and given the symbol μ_0). Unless light were an electromagnetic wave, it is impossible to see why there should be any connection at all between the speed of light and these rather abstract electrical quantities. Yet, the relation predicted by Maxwell ($c^2 = 1/\epsilon_0\mu_0$) correctly predicts the speed of light.

A modern value for the speed of light is 299,792,458 metres per second (or, if you prefer old-fashioned units, this is approximately 186,282 miles per second). This means that the light reaching Earth from the Moon arrives just over a second after it was emitted. Sunlight is just over 8 minutes old by the time it reaches Earth. Light from Jupiter arrives at Earth somewhere between just over half an hour and the best part of an hour after it was emitted, all depending on the relative position of the two planets around their orbits (a difference that allowed Rømer to do his measurement in 1676). Light is extraordinarily fast, but its speed is set by properties which are all to do with electromagnetism.

Maxwell's predictions were vindicated by the experiments of Heinrich Hertz, who in 1886 showed that an electromagnetic wave generated by the discharge across a spark gap could be picked up by a receiver consisting of a copper wire and a brass sphere placed close by. Subsequent measurements showed that the speed of these electrically generated waves was precisely the speed of light. Hertz had essentially built the first radio transmitter and receiver, though his early death in 1894 aged 36 robbed him of seeing his invention developed by others such as Guglielmo Marconi. Hertz hadn't foreseen the practical possibilities of radio, but he was in good company. William Thomson, who became Lord Kelvin in 1892, famously puffed 'Radio has no future'. Kelvin's track record on futurology was poor. He was similarly dismissive about X-rays and aviation.

While at King's College London, Maxwell chaired a committee to decide on a new system of units to incorporate this new understanding of the link between electricity and magnetism (and which became known as the 'Gaussian', or cgs (centimetre-gram-second), system – though 'Maxwellian system' would have been more appropriate). In 1865, he resigned his chair at King's and moved full time to Glenair, where he wrote his *Theory of Heat* which introduced what are now known as Maxwell relations and the concept of 'Maxwell's demon'. He applied for, but did not get, the position of Principal of St Andrews University, but in 1871 was appointed to the newly established Professorship of Experimental Physics in Cambridge (after William Thomson and Hermann von Helmholtz both turned the job down). There he supervised the building of the Cavendish Laboratory and wrote his celebrated *A Treatise on Electricity and Magnetism* (1873) in which his four electromagnetic equations first appear. In 1877, he was diagnosed with abdominal cancer, and he died in Cambridge in 1879 aged 48. In his short life, Maxwell had been one of the most prolific, inspirational, and creative scientists who has ever lived. His work has had far-reaching implications in much of physics, not just in electromagnetism. He had also lived a devout and contemplative life, remarkably free of ego and selfishness, and was generous and courteous to everyone. The doctor who tended him in his last days wrote:

> I must say that he is one of the best men I have ever met, and a greater merit than his scientific achievements is his being, so far as human judgement can discern, a most perfect example of a Christian gentleman.

Maxwell summed up his own philosophy as follows:

> Happy is the man who can recognize in the work of Today a connected portion of the work of life, and an embodiment of the work of Eternity. The foundations of his confidence are unchangeable, for he has been made a partaker of Infinity.

Maxwell was the first person to really understand that a beam of light consists of electric and magnetic oscillations propagating together. The electric oscillation is in one plane, at right angles to the magnetic oscillation. Both of them are in directions at right angles to the direction of propagation. Thus if you could look at a beam of light, you would see the structure as shown in Figure 8. The oscillations of electricity and magnetism in a beam of light are governed by Maxwell's four beautiful equations, operating together like the cogs, wheels, and spindles inside an intricate machine, each playing a role to keep the whole wonderful mechanism in perfect harmony.

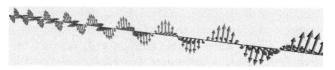

8. **An electromagnetic wave. The electric field oscillates in one plane and the magnetic field oscillates in a plane perpendicular to it**

Maxwell's discovery of electromagnetic radiation was extraordinarily far-reaching. Though the speed of the waves is fixed, the wavelength (the distance between the peaks and troughs of the wave motion) can take any value. The wavelength can be tens, hundreds, or thousands of metres, and then the waves are called radio waves. At the centimetre level, they are called microwaves (the ones used in a microwave oven usually have a wavelength of around 12 centimetres). When the wavelength is well below a millimetre and runs down to just below a micrometre, the radiation is called infrared (and is the sort of electromagnetic radiation emitted by objects close to room temperature, so that infrared cameras are used for thermal imaging). The visible band of the spectrum, what we usually call 'light', is crammed into a narrow region between about 0.4 and 0.7 micrometres. Smaller than that, the radiation is ultraviolet (contributing to your suntan and/or sunburn), and below 10 nanometres they are called X-rays (or gamma-rays if the wavelength is shorter than one-hundredth of a nanometre).

The perspective provided by Maxwell's discovery was a lofty one. Many of the discoveries of new phenomena that were made in the 19th century – the X-rays that provided pictures of the skeleton inside a body, the radio waves that could be used to transmit messages, infrared radiation that relates to the transfer of heat – all were examples of electromagnetic radiation, nothing more or less than vibrations of electric and magnetic fields. Science was clearly doing what it does best, unifying disparate and seemingly unconnected phenomena and demonstrating that they originate from a single physical cause. No wonder that many spoke of the imminent end of science, a point soon to be reached whereby all the relevant scientific questions would be answered and the final picture would be encompassed in a single theory. But just as everything seemed to be on the verge of resolution, it all started to fall apart. The spark that ignited the new revolution was nothing more than Maxwell's own beautiful equations.

Chapter 5

Magnetism and relativity

The aether

If light is a wave, as Maxwell had argued, then it must propagate through a medium. Sound moves through air, water waves move through the ocean, so light must move through something. How can you have a wave without a medium through which it moves? But what is there in the empty wastes of space through which light travels to us from distant stars? No substance could be detected, and as far as anyone knew, space was completely empty. But the dogma asserted that a wave must propagate through a medium and so there must be one. Therefore scientists postulated the existence of a medium and gave it a name: the luminiferous aether.

This mythical medium had to have several properties. It had to be transparent, so light could go through it. It had to be very rigid so that it could support the very high-frequency waves of light. It had to fill all space everywhere since light travels all over the place. It had to have no mass, since it doesn't appear to weigh anything down and the planets orbit happily around the Sun without slowing down, even though they are flowing through the aether, so it can't give rise to any viscous drag. The luminiferous aether would be strange stuff indeed.

Albert Einstein was ultimately responsible for getting rid of the aether, as we shall see later, but the credit should also be shared with Albert Michelson and Edward Morley who performed an experiment (somewhat inspired by a proposal by Maxwell) which aimed to detect the aether. The idea was that if the aether is somehow fixed in space, and planet Earth itself is hurtling through space in its orbit around the Sun, then the Earth is moving with respect to the aether. If you measure the speed of light (fixed with respect to the aether, but not with respect to the moving Earth), it should be different depending on whether you measure it along the direction of the Earth's motion through space or perpendicular to it. Michelson and Morley consequently attempted to measure this effect on the speed of light in a device called an interferometer which was either aligned along or perpendicular to the Earth's motion through the aether; no effect was found, casting some doubt on the existence of the aether and/or implying that new physics needed to be proposed.

There were various get-outs. It was a small effect and maybe the experiment was insufficiently sensitive? That was quickly ruled out. Maybe the interferometer shrunk in particular directions with respect to the aether wind. If you posited a certain degree of shrinkage, you could get the numbers to come out right, but it was a fudge. Maybe the Earth dragged the aether around with it so that near to the Earth a thick viscous envelope of aether clings to the planet like syrup to a spoon? Then the Earth would not be moving with respect to the adhering aether and so the Michelson–Morley experiment would see nothing. But aether drag could be ruled out by astronomical observations of stellar abberation, and so this was also a non-starter. Something was very wrong.

The speed of light is absolute

The Michelson–Morley experiment predated Einstein's theory of relativity, but it was not the main motivation for his revolutionizing work. In fashioning his theory of relativity, Albert

Einstein was guided not so much by the results of this complicated experiment but by a more fundamental desire to construct a principle-based theory, a construction which had as its foundation some grand statement which signified something important about the Universe.

Above all, Einstein's work on relativity was motivated by a desire to preserve the integrity of Maxwell's equations at all costs. The problem was this: Maxwell had derived a beautiful expression for the speed of light, but the speed of light with respect to whom? If you drive a car at 70 miles an hour and switch your headlights on, then should you measure the speed of light from your headlights with respect to you, the driver, or with respect to someone standing by the side of the road? You would naively think that the two answers would differ by 70 miles an hour. In that case, which one of those two answers would agree with Maxwell's beautiful expression? Worse was to follow, since it was realized that if Maxwell's equations work according to one observer, then they do not work for a second observer who is moving at a fixed speed with respect to a first.

Einstein deduced that the way to fix this would be to say that all observers will measure the speed of any beam of light to be the same. However fast the observers are moving with respect to each other, and in whatever direction, they will all measure the speed of the same beam of light to be exactly the value that Maxwell had calculated. By holding fast to the constancy of the speed of light with respect to everyone, Einstein forced common sense to bend round some strange corners to compensate.

Albert Einstein

Einstein's academic career had not got off to a racing start. In 1895, he failed to get into the prestigious Eidgenössische Technische Hochschule (ETH) in Zürich, and was sent to nearby Aarau to finish secondary school. He enrolled at ETH the following year, but after his degree he failed to get a teaching assistant job there. Instead Einstein began teaching maths at technical schools in Winterthur and Schaffhausen, finally landing a job at a patent office in Bern in 1902 where he was to stay for seven years. Though Einstein was physically present in the office, fulfilling his undemanding role as 'technical examiner third class', his mind was elsewhere and he combined the day job with doctoral studies at the University of Zürich.

In 1905, this unknown patent clerk submitted his doctoral thesis (which derived a relationship between diffusion and frictional forces, and contained a new method to determine molecular radii) and also published four revolutionary papers in the journal *Annalen der Physik*. The first paper proposed that Planck's energy quanta were real entities and would show up in the photoelectric effect, work for which he was awarded the 1921 Nobel Prize. The citation stated that the prize was 'for his services to Theoretical Physics, and especially for his discovery of the law of the photoelectric effect'. The second paper explained Brownian motion on the basis of statistical mechanical fluctuations of atoms. The third and fourth papers introduced his special theory of relativity and his famous equation $E = mc^2$. Any one of these developments alone was sufficient to earn him a major place in the history of physics; the combined achievement led to more modest immediate rewards: the following year, Einstein was promoted by the patent office to 'technical examiner second class'.

Living in Switzerland, Einstein was familiar with a well-run train service. I am a fairly frequent traveller on Swiss trains and am always impressed with their smooth ride and punctual operation. Sitting at a station, looking out of the train window, and seeing the neighbouring train move with respect to you, it can be very hard to work out whether your train is stationary and the other train is moving backwards, or your train is gliding forwards and the other train is stationary. Which is correct? The presence of the station gives us a frame of reference so that we can settle the question of which train is moving. But what if there were no station?

Einstein had begun to think about the implications of this observation which implies that there is no absolute motion and all motion is relative. Whichever 'frame of reference' you find yourself in, you should be able to come up with a consistent explanation of the Universe around you based upon physical laws. Einstein had realized that the physical laws formulated up until his time did not fit with this idea and his theory of relativity provided the correct alternative.

Something's got to give

Requiring the speed of light to be the same for all observers leads to some unexpected consequences. For a start, if you have two events which are thought to be simultaneous, such as a clock in London striking midnight at exactly the same moment as a clock in New York strikes seven o'clock in the evening (the two locations are five time zones apart), then an observer moving in a spaceship at high speed with respect to the Earth will deduce that one of those events comes before the other (which one comes first depends on which direction the spaceship is travelling). We are so used to thinking about the notion of something in one place happening at 'exactly the same time' as something happening in another place, that we don't realize that such a statement is not universal, in the sense that not all observers would agree on it. This is just one of the bizarre things that happen when you insist that the speed of light is constant for all observers.

But there's more. If an astronaut travels in a spaceship in uniform motion at some speed with respect to you, the astronaut's watch will run slower than yours. For her, time slows down with respect to you. However, she will deduce the same about you and assume it is your watch that has slowed down. This extraordinary effect is called *time dilation*, and it is observed in the laboratory.

Short-lived radioactive particles which are set in motion in particle accelerators are observed to live longer than when they are stationary; their internal clocks have slowed down and this results in them getting further along the beamline than you would expect before they decay. Uniformly moving objects are also found to shrink in the direction of their motion, an effect called *length contraction*. In fact, the radioactive particles in the experiment referred to above would observe themselves to be stationary and their surroundings moving towards them. From their perspective (imagine for a moment that they have one), it is not they have lived longer before decaying, it is that their surroundings have shrunk because of length contraction. Thus all observers can find interpretations for their particular frame of reference based on Einstein's theory.

Why were these strange effects not noticed before? Einstein showed that the effects of time dilation and length contraction only become apparent when the velocity of an object approaches the speed of light. The fastest humans have ever travelled was on one of the Apollo flights to the Moon and then, at around 11 kilometres per second, this was less that 0.4% of the speed of light. Even at these speeds, relativity only gives a small correction to what you would assume using pre-relativity physics. For example, on a flight from London to Washington, the effect of time dilation means that your watch would run slow by about 10 to 20 nanoseconds. Airport delays are typically much longer than this. (In fact, because you fly at altitude, you fly in a slightly weaker gravitational field, and this appears to speed up your watch compared to that of a stationary observer by about 50 to 60 nanoseconds due to Einstein's general theory of relativity, but that's another story.)

Relativity and magnetism

Relativistic effects look pretty small and ignorable for general life, and this not a book about relativity (and if you want to know more, please consult *A Very Short Introduction to Relativity* by Russell Stannard). However, what is important for our story is that Einstein showed that magnetism is a purely relativistic effect, something that wouldn't even be there without relativity. Magnetism is an example of relativity in everyday life.

Imagine a Universe populated only by stationary electrical charges. Nothing moves and nothing happens. The charges sit in space and lines of electric field emanate from each positive charge and converge into each negative charge. Now imagine viewing such a Universe from a spaceship travelling at some speed with respect to the fixed charges. From this point of view, the charges are now moving with respect to your spaceship. Einstein's equations show that from your perspective some of the electric field is transformed into magnetic field. Magnetic fields are what electric fields look like when you are moving with respect to the charges that 'cause' them.

When you think about it, every time a magnetic field appears in nature, it is because a charge is moving with respect to the observer. Charge flows down a wire to make an electric current and this produces magnetic field. Electrons orbit an atom and this 'orbital' motion produces a magnetic field. As we will see in Chapter 9, the magnetism of the Earth is due to electrical currents deep inside the planet. Motion is the key in each and every case, and magnetic fields are the evidence that charge is on the move. Just as Einstein, sitting on a train in a Swiss station, notices only the relative motion of the adjacent train with respect to him, the relative motion of an electrical charge is perceived by an observer via a magnetic field.

The remarkable thing about magnetism as a relativistic effect is that ordinary electrical currents do not move very fast. If you work out the speed of the charges in a wire, the so-called *drift velocity*, you end up with a small number, perhaps a few millimetres per second. This is clearly much, much less than the speed of light, and so why do you ever notice a magnetic field?

A current-carrying wire contains very large numbers of positive and negative charges. The positive charges are fixed (in the centres of the atoms) and the negative charges (the electrons) are mobile and flow along the wire. However, the number of positive charges equals the number of negative charges and so the two sets of charges completely cancel out. Therefore, the wire produces no electric field because it has no net electrical charge. This is why your headphone leads don't attract nearby objects to them. Electric forces are incredibly strong; they are what hold all rigid objects together, bridges, walls, and human beings; electric forces stop you falling through the floor. But they all cancel out in a current-carrying wire because all the positive and negative charges balance. However, the tiny relativistic effect, what we call the magnetic field, which is due to the *motion* of the negatively charged mobile electrons, remains and is not cancelled out.

To make this even more concrete, let's put in some numbers. Imagine you have two wires, each carrying current in the same direction, and let's say the drift velocity in each wire is three millimetres a second, which is a million millionth of the speed of light (i.e. $10^{-12}c$ where c is the speed of light). If the wires contained no compensating positive charges, the force between the negatively charged electrons which carry the current would be electrostatic and very strongly repulsive ('like charges repel'). But the compensating charges are there and so the electrostatic forces are zero. The magnetic attractive force between the two wires is about a factor of 10^{24} weaker than the uncompensated electrostatic force, because it is just a tiny relativistic correction, but Ampère was nevertheless able to observe it in his experiments because the positive and negative charges really do cancel completely, and the enormous electrical effect vanishes leaving only the tiny relativistic effect remaining.

Einstein's theory of relativity casts magnetism in a new light. Magnetic fields are a relativistic correction which you observe when charges move relative to you. But in this chapter, we have only really been able to think about magnetic fields, and not about the magnetic materials that can produce them. Why is lodestone spontaneously magnetic? To answer that question, we must turn once more to the first decades of the 20th century when another revolution in physics was taking place, and once again magnetism would be centre stage.

Chapter 6
Quantum magnetism

What keeps a magnet working? Pierre Curie (Marie's husband) beavered away in his lab to try and answer this question, measuring the properties of magnetic substances at different temperatures and in different magnetic fields. He showed that in many seemingly non-magnetic materials the applied magnetic field tends to make the material more magnetic, but that warming the material reduces its magnetism. Each atom in the material behaves like a tiny magnet, and Curie deduced that a magnetic field lines up the atomic magnets, but heat randomizes them. Curie showed that materials become more susceptible to an applied magnetic field when you cool them down, because the randomization effect is smaller (this effect is behind what is now known as Curie's law).

He also studied compounds which, like lodestone, are spontaneously magnetic even when you don't put them in a magnetic field. They are known collectively as ferromagnets. He showed that above a certain critical temperature, now known in his honour as the Curie temperature, ferromagnets lose their magnetism. For iron, the Curie temperature is 770°C, for lodestone (magnetite, chemical formula Fe_3O_4) it is 585°C. Pierre Weiss, another French physicist, took Curie's formula for non-ferromagnetic materials and tried to understand ferromagnets with it. He proposed that ferromagnets contained within them their own internal magnetic field which forced the atomic magnets to line up spontaneously. Weiss's idea was clever but contained a flaw. The size of Weiss's supposed internal magnetic field came out to be ridiculously big, a thousand times larger than is observed close to any piece of iron. A full explanation of the phenomenon of magnetism would only come when the new science of quantum mechanics had been developed.

Let's step back a bit. Since Oersted's work, it was clear that you could produce magnetic fields in two different ways: by using a magnetic material (e.g. the lodestone) or by using an electric current flowing along a wire. However, there was a difference between these two methods. The electric current has to be driven, powered by a battery (which will not last forever), and the wire will get warm (as all current-carrying wires do to a varying extent, an effect that is put to good use in an electric toaster). The puzzle is the following: the lodestone has no external battery powering it (does it therefore have some kind of internal power source?) and it doesn't get hot (so does that mean the currents inside it are very different from the ones that flow down cables?). In the 19th century, the principle of conservation of energy had begun to be articulated, and so this everlasting nature of the magnetism in ferromagnets such as lodestone came to be appreciated as even more remarkable.

Atoms in lodestone contain electrical currents due to the electrons which orbit the nucleus and these atomic currents are indeed rather special. They are more similar to the current in superconducting wires than to the current flowing in a copper wire. A coil of superconducting wire can carry a circulating current without a power source and without dissipating heat, and generates an intense magnetic field (such superconducting coils are used in MRI scanners, familiar in many hospitals). The current flowing around an atom also travels without dissipating heat because it originates from the stable orbits of the electrons around the nucleus. This can only be rationalized using the counter-intuitive logic of quantum mechanics which allocates to each electron a well-defined energetic state. To really explain magnetism, we have to enter the quantum world.

Quantum mechanics

Quantum mechanics revolutionized physics. The distinction between waves and particles was abandoned. The fundamental description of the Universe became probabilistic. It was realized that various quantities, such as angular momentum, could only be changed by discrete amounts. Particular pairs of quantities, such as the position and momentum of a particle, could not be exactly known simultaneously, and the increasing precision to which one quantity was determined led to a decreasing precision in the determination of the other (Heisenberg's famous uncertainty principle). The development of quantum mechanics in the 1920s motivated physicists to tackle all the unsolved problems of physics with the new methods and see if they worked (they mostly did). But what was the evidence for any of this new way of thinking?

The evidence that was persuasive at the time was a number of rather abstract physics experiments concerning the nature of atomic spectra or the interaction between light and metal surfaces. Each was important in its own way, but what ought to have played an important role in retrospect was something far, far simpler: the observation that magnets work.

The crucial step was made by an unknown Dutch scientist called Hendreka van Leeuwen, and what she showed was that magnets couldn't exist if you just use classical (i.e. pre-quantum) physics. Hendreka van Leeuwen's doctoral work in Leiden was done under the supervision of Lenz and the work was published in the *Journal de Physique et le Radium* in 1921. Unfortunately, it subsequently transpired that her main result had been anticipated by Niels

Bohr, the father of quantum mechanics, but as it had only appeared in his 1911 diploma thesis, written in Danish, it was unsurprising she hadn't known about it. Their contribution, though conceived independently, is now known as the Bohr–van Leeuwen theorem, which states that if you assume nothing more than classical physics, and then go on to model a material as a system of electrical charges, then you can show that the system can have no net magnetization; in other words, it will not be magnetic. Simply put, there are no lodestones in a purely classical Universe.

This should have been a revolutionary and astonishing result, but it wasn't, principally because it came about 20 years too late to knock everyone's socks off. By 1921, the initial premise of the Bohr–van Leeuwen theorem, the correctness of classical physics, was known to be wrong: the physical Universe is a quantum one and it's not surprising that a calculation assuming classical physics gives you the wrong answer. But when you think about it now, the Bohr–van Leeuwen theorem gives an extraordinary demonstration of the failure of classical physics. Just by sticking a magnet to the door of your refrigerator, you have demonstrated that the Universe is not governed by classical physics. The need for quantum theory is often presented by describing somewhat esoteric experiments in the physics laboratory: the photoelectric effect, the detailed appearance of atomic spectra, or the bouncing of electrons off a crystal. All these were indeed important historically for highlighting the failure of classical physics to provide an adequate explanation of the world, but none of them are child's play to do. It's rather charming to realize that the same need for quantum theory can be shown simply by playing with magnets.

Understanding real materials

In the 1920s and 1930s, physicists took up the challenge to apply the new quantum mechanics to magnetism, finding that the magnetic properties of many real materials could be explained by quantum mechanics. For example, it is known that most real substances are weakly diamagnetic, meaning that when placed in a magnetic field they become weakly magnetic in the opposite direction to the field. Water does this, and since animals are mostly water, it applies to them. This is the basis of Andre Geim's levitating frog experiment: a live frog is placed in a strong magnetic field and because of its diamagnetism it becomes weakly magnetic. In the experiment, a non-uniformity of the magnetic field induces a force on the frog's induced magnetism and, hey presto, the frog levitates in mid-air. A frog was chosen for this experiment because it was small enough to fit into the magnet, but with a big enough magnet it would in principle be possible to make pigs fly! Quantum mechanics nicely explains diamagnetism, and also the related phenomenon of paramagnetism in which a material becomes magnetic in the same direction of the magnetic field applied to it. Crystals of copper sulphate (the bright blue crystals that children often grow at school) are good examples of paramagnets.

However, the really exciting problem to get one's head around is ferromagnetism, the magnetism of lodestone. It is a much larger effect and proved more resistant to analysis. The solution came from thinking about a very odd symmetry of quantum mechanics. Quantum mechanics states that the fundamental description of a quantum object, such as an electron, or a set of electrons, is controlled by a special mathematical function called the *wave function*. The wave function varies in space, and the magnitude of the square of this function at a particular place gives you the probability that the quantum object is found there.

Now take two identical particles located at two different places in space, and swap them. What you have produced has to be the same as you started with, doesn't it? The two particles are absolutely identical and so the situation you've ended up with must be the same as you started with. We now know that for some of the particles in the Universe (known as *bosons*), this is completely true. However, for the other particles in the Universe (known as *fermions*), something rather special happens: the quantum mechanical wave function describing the two particles changes sign. This seems very strange, but physical reality only depends on the square of the wave function and so reality is not affected by this sign change. But the change of sign is there and does have consequences.

Electrons are fermions and so they do this sign-change trick when you swap them around. Let's think about a piece of iron and focus in on two iron atoms inside that piece. Think about two electrons, one associated with a particular iron atom and another associated with a neighbouring iron atom. We want to consider the wave function describing those two electrons. We know now that when we swap the two electrons, the wave function must change sign. However, the wave function describes various properties of the two electrons, such as their position in space and their magnetic orientation (a property called 'spin', which is discussed in more detail in the next chapter). Which bit changes sign when you swap the two electrons? Is it the bit that is associated with position or the bit associated with spin? In fact, it transpires that it is either one or the other (but it cannot be both; two sign changes give you no net sign change).

It turns out that in materials like iron you can save a lot of energy when it is the position bit that changes sign when you swap the two electrons. This is because in this configuration the two electrons avoid each other quite efficiently, thereby minimizing the energy cost of electrostatic repulsion. In this case, the spin bit of the wave function doesn't change sign, and the configuration that satisfies this is when the two spins are aligned. This is the mechanism for keeping the spins aligned in a ferromagnet (Figure 9(a)). Because this mechanism involves this strange property of what happens to two electrons when you swap them, the property causing the spins to align in a ferromagnet is known as the *exchange interaction*.

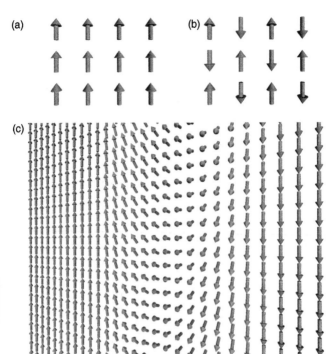

9. (a) A ferromagnet; (b) an antiferromagnet; (c) a domain wall

This way of thinking explains an important fact. In a material like iron, the magnetism persists up to very high temperatures, up to the Curie temperature, which is 770°C, showing that the interaction is very strong. This strength can be traced back to the large energy associated with electrostatic repulsion.

If all this is true, why isn't every piece of iron you come across magnetic? The answer to this question is that every piece of iron *is* magnetic, but the magnetic structure often breaks up into small regions called domains. In each domain, the magnetism in every atom is aligned in the same way (we will again refer to atomic magnets as *spins*). Thus a domain contains spins that are aligned and point in the same direction, but different domains have all the spins within them that aligned differently. This means that, from the outside, the piece of iron does not appear to be magnetic because the effect of all the individual domains cancel out.

Why do they do that? This behaviour comes from an attempt to minimize energy. If a piece of iron exists as one domain, then it will produce a magnetic field outside which fills the space around it. But this magnetic field costs energy. Therefore, it is favourable for the magnetic structure to break up into domains because this removes the outside stray field and hence saves energy. Now even doing that comes at a price, because in the region between the domains we have to have what is called a domain wall (see Figure 9(c)). In a domain wall, the spins twist round from the configuration in one domain to that in its neighbour. This twisting also costs energy. So whether you get a single domain or many domains depends on a subtle balance between energy costs.

In a permanent magnet, like a piece of lodestone, or the magnet inside a motor, dynamo, or at the back of a loudspeaker, it is easy for the magnet to be in a single domain state. In such magnets, the cost of a domain wall is considerable and so they do not tend to form easily. If they do form, they are usually stuck in one place and very difficult to move. Such magnets keep their magnetism (unless they are warmed up above their Curie temperature), and to reflect their stubborn and unperturbable nature they are called hard magnets.

However, a piece of pure iron is a soft magnet in which the cost of making a domain wall is very small. Such magnets easily break up into a multi-domain structure. They can be easily magnetized and just as easily demagnetized. This makes them very useful in applications in which the magnetization needs to be switched on and off. For example, in the core of a transformer, there is a piece of iron which is magnetized backwards and forwards 50 or 60 times a second (depending on whether your electricity runs at 50 or 60 Hz), and one wants them to do this as easily as possible so that very little heat is generated. Soft magnets are ideal for this because the domain walls can move through them very easily.

Antiferromagnetism

We have seen that in some materials the exchange interaction forces neighbouring spins to be parallel and the resulting material is a ferromagnet. There are occasions when the exchange interaction favours antiparallel alignment and, in this case, the material becomes an *antiferromagnet*, in which neighbouring spins are arranged antiparallel to each other as shown in (Figure 9(b)). This idea was first suggested by the French physicist Louis Néel in the 1930s, and the configuration is usually referred to as the Néel state.

Quantum mechanics has an extra trick up its sleeve for antiferromagnets. Consider two neighbouring spins. If the interaction between them is antiferromagnetic, then one possible configuration could be considered as 'up down' (which is code for: the first spin is up and the second spin is down). However, another possible configuration is 'down up' (which is code for: the first spin is down and the second spin is up). Quantum mechanics possesses the curious property that it allows reality to consist of a combination of unrealized possibilities, such as Schrödinger's unfortunate (and imaginary) cat which is simultaneously both fully alive and fully dead. Thus, for our two spins, both 'up down' and 'down up' are realized together and the ground state should actually be written:

Wave function = up down − down up

This is known as a singlet state (because there is only one way of doing it), and is a particular combination of the two configurations considered above (it ends up being necessary to write down the difference of the two configurations, rather than the sum). This means that it is not possible to know the state of either spin, only that whatever the first spin is, it is antiparallel to the second spin.

We have been using the word 'spin' as a shorthand for the magnetism of an individual atom or particle. In the following chapter, we will describe what spin is and how it was discovered.

Chapter 7
Spin

In the early days of quantum mechanics, back in the 1920s, the word 'spin' began to be used to describe a strange new property of the electron connected with its intrinsic magnetism. It was first thought that the magnetic properties of the electron were due to it spinning on its own axis, rather like a basketball spinning on the finger of a Harlem Globetrotter, and hence the name 'spin' seemed entirely appropriate. However, as an electron was also known to be a point particle, this concept makes no sense. How can a vanishingly small point rotate?

Let's step back a bit to the experimental evidence for the spin of an electron before worrying about what it actually is. It all started with a quest to understand how atomic vapours glow when you heat them or subject them to an electrical discharge. The kind of light emitted from such a vapour tells a story concerning the electrons inside each atom and the path each of them takes in their orbits around the nucleus. Quantum theory shows that the orbits are not random, but that the electrons are fixed in a small set of allowed orbits, each one of which is associated with a fixed value of energy. Light can be emitted when an electron transfers from one orbit to another. The energy of the emitted light makes up for the difference between the energies in the two orbits.

Physicists often choose to forget about the details of the orbits and think simply about the electrons in an atom occupying particular *energy levels*, although these energy levels are each associated with particular orbits. Sodium street lamps give a familiar orange glow originating from a particular atomic transition in sodium, an electron moving from one particular energy level to another, resulting in the emission of a photon with an energy equal to the difference in energy between the energies in the upper and lower levels. This well-defined frequency of emitted light shows up as a single line in the spectrum of sodium, and as such is called an *emission line*. However, close examination of this emission line reveals it to be split into two. This seemed to suggest that one of the energy levels was actually also split into two very closely spaced levels. This was a first clue that there was something two-valued about the electron in sodium (the physicist Wolfgang Pauli called it a 'two-valued quantum degree of freedom').

A further experiment can be done with the emission of light from sodium. A magnetic field can be applied to the sodium vapour to see what happens to the emission lines. It is found that a magnetic field causes the emission lines to change frequencies because different energy levels shift by varying amounts in a magnetic field. This effect results from the magnetic field interacting with the orbits of electrons around the atom, and the process is known as the *Zeeman effect*, in honour of the Dutch physicist Pieter Zeeman who first tried it out. Quantum mechanics forces the orbits of electrons around the atom to take certain fixed configurations with certain allowed speeds of rotation. It turns out that this is done in such a way that the angular momentum of the electron takes integer (i.e. whole number) values (when measured in the appropriate units, given by Planck's constant $\hbar$). These angular momentum states all have the same energy and so, without a magnetic field present, they are hidden within the same emission line. However, the magnetic field causes these different angular momentum states to separate in energy and so the resulting emission lines split into a series of closely-spaced new lines. These angular momentum states of an atom were already well known but in some atoms extra transitions were noticed which could not be explained by the orbits of the electron around the nucleus (this was dubbed the anomalous Zeeman effect because the observations didn't fit in with the then current picture). Again, these extra transitions pointed to some extra degree of freedom.

These effects in atomic spectra seem rather obscure and technical, but in the early 1920s they clearly demonstrated that all had not been understood. In the theory that everyone used, the electron energy levels were labelled with three quantum numbers (called principal, azimuthal, and magnetic, and given the symbols n, l, and m_l) and these took integer values and followed certain rules. They could be derived from Schrödinger's equation, related to properties of the orbits of electrons around the atom (via energy and angular momentum) and explained most of the features in atomic spectra, but not all. Wolfgang Pauli deduced that another quantum number was required which described the 'strange two-valuedness' of the properties of the electron 'which cannot be described classically'. He refrained from making any interpretation as to what this extra property might be.

In 1925, a 21-year-old physicist called Ralph Kronig proposed that there could be an extra source of angular momentum in the atom not yet accounted for. Yes, electrons orbit around the nucleus, but what about the self-rotation of the electron itself? Could this be the origin of the strange effects in atomic spectra? Wolfgang Pauli positively hated the idea. The electron was known to be exceptionally tiny, possibly even point-like. If it were rotating on its own axis, the velocity at its surface would greatly exceed that of light, violating the theory of relativity. When Kronig met Pauli and discussed his idea, Pauli was cool and unenthusiastic. Kronig decided not to publish.

In September of that same year, two physicists, George Uhlenbeck and Samuel Goudsmit, came up with essentially the same idea as Kronig. Goudsmit knew a lot about atomic spectra and was able to educate Uhlenbeck on the latest ideas. Uhlenbeck had the virtue of ignorance of the subject which meant that he asked Goudsmit lots of innocent-sounding but rather relevant questions. When hearing that the traditional integer quantum number scheme (the n, l, and m_l) did not explain the atomic spectra, he suggested that they try half-integer quantum numbers. (It works out that if you give the self-rotation of an electron the quantum number one-half, then it naturally gives you two-valuedness because the possible values a measurement of the intrinsic angular momentum of the electron can give you are plus or minus one-half.) Uhlenbeck later recalled: 'It was then that it occurred to me that, since (as I had learned) each quantum number corresponds to a degree of freedom of the electron, the fourth quantum number must mean that the electron had an additional degree of freedom – in other words the electron must be rotating!' They fired off a paper to the journal *Nature* entitled 'Spinning electrons and the structure of spectra'.

A factor of two

However, one embarrassing feature of Uhlenbeck and Goudsmit's theoretical approach was that it could be used to make a precise prediction of the energy of the splitting of the energy levels in hydrogen due to the interaction between the electron and the nucleus and their prediction was out by a factor of two. Now physics students not infrequently drop factors of two or lose minus signs due to algebraic sloppiness, forcing them to spend considerable effort to trace their mistake. It happens to professional physicists too, and sometimes a silly error (such as mixing up inches and millimetres) slips through and can, for example, be enough to cause a spacecraft to crash into Mars rather than land gracefully on its surface.

For Goudsmit and Uhlenbeck, the missing factor of two was not the result of a mistake. In fact, they were initially completely unaware that the answer they had worked out was wrong. Very shortly after publication, they received a letter from Werner Heisenberg congratulating them on their 'brave note' and enquiring how they had got rid of the factor of two? They immediately did the calculation and discovered to their horror that Heisenberg was right. Kronig in fact had earlier done the same calculation in his model and also found that the theory comes out wrong by a factor of two; this was another reason why he declined to publish. Goudsmit and Uhlenbeck tried to withdraw their paper, but it was too late, and 'Spinning electrons and the structure of spectra' was published in February 1926. Kronig, probably annoyed that he had had the same idea and failed to publish, sent off a critical letter to *Nature*, which was published a few months later. He laid into Goudsmit and Uhlenbeck's result, showing that their assumption of a spinning electron created more problems than it solved, concluding somewhat sourly:

> The new hypothesis, therefore, appears rather to effect the removal of the family ghost from the basement to the sub-basement, instead of expelling it definitely from the house.

The problem of the mysterious factor of two was cleared up by Llewellyn Thomas in an article published (again in *Nature*) in April 1926. Thomas did a rather ingenious and sophisticated calculation involving Einstein's theory of relativity which took into account the transformation into the reference frame of the rotating electron; satisfyingly, it explained precisely the missing factor of two. Thomas concluded that the 'interpretation of the fine structure of the hydrogen lines proposed by Messrs. Uhlenbeck and Goudsmit now no longer involves any discrepancy'. Uhlenbeck and Goudsmit could breathe a sigh of relief.

Otto Stern and Walter Gerlach

Something of the peculiarity of electron spin was shown in an experiment performed by Otto Stern and Walther Gerlach in Frankfurt in 1922. Stern was an assistant to the great Max Born, one of the founding fathers of quantum mechanics. The idea for the experiment came to Stern while he was lying in bed. He mused on the fact that an electron orbiting an atom is a circulating current and hence the atom can behave like a little magnet. The electron could then feel a force if placed in a magnetic field gradient, that is a magnetic field that varies with position. Stern discussed his idea with Born, who was rather doubtful about it, but Stern decided to have a go at testing it out anyway and recruited Walther Gerlach, who was working at a neighbouring institute, to lend a hand.

They decided to choose silver atoms to perform the experiment. Silver was heated in an oven to a high temperature so that it became a vapour, and the hot vapour was allowed to pass out of the oven through a couple of thin slits into an evacuated region to produce a collimated beam of silver atoms. The beam was then passed through a magnetic field gradient (made simply by constructing a magnet in which the north pole has a very diffent shape from the south pole). The beam of silver atoms then passed onto a glass slide. After running the experiment for a while, they removed the glass slide to have a look to see where the silver landed and thus infer how the beam of silver atoms had been affected by the magnetic field gradient.

The experiment was difficult and a bit temperamental, and they could not run the atomic silver beam for very long. They thus managed to deposit only a rather feeble quantity of silver on the glass slide. Disappointingly, the glass slide seemed to show no trace of silver. However, after Otto Stern had accidentally puffed his cheap cigar smoke all over it (he was particularly fond of cheap cigars), the pattern suddenly and magically emerged. It seems that the cheap cigar smoke must have contained a lot of sulphur and turned the very thin layer of silver deposited on the slide into jet-black silver sulphide, which was then much more easily visible. These days, a sensitive atomic beam experiment would be carried out by people wearing special suits and protective head coverings in a dust-free air-conditioned clean room, but in the 1920s it was *de rigueur* for the experimenters to be clad in tweed jackets and be perpetually wandering round their dirty lab surrounded by huge clouds of pipe smoke. On this particular occasion, that helped.

The experiments were nevertheless painstaking and in the difficult financial environment of postwar Germany, the lab was running out of money. Fortunately, help was at hand after Born wrote a begging letter to Henry Goldman in New York. Goldman was a founder of the investment firm Goldman Sachs, but had family roots in Frankfurt and his cheque bankrolled the Stern–Gerlach experiments.

If classical physics had been right, then the silver atoms in the gas would have been arranged randomly and the action of the field gradient would have simply been to blur out the silver trace on the glass slide. Some of the atoms would have been deflected up, some down, some not at all, and everything in between. But what Stern and Gerlach found was truly staggering. The beam split into two. Half the atoms were deflected up, half of them were deflected down.

Stern and Gerlach didn't realize it at the time, but there is nothing particularly special about silver. It is the outermost electron in silver that is responsible for the effect, and in fact the experiment was repeated five years later with hydrogen, demonstrating that silver is not a vital component. In fact, if the experiment were to be done with a simple beam of electrons, the effect would, in principle, be the same (though the experiment is more complicated because the electrons are charged; the great thing about silver atoms and hydrogen atoms for this experiment is that they are electrically neutral). For the sake of simplicity of description, we will discuss Stern and Gerlach's result as if we are just studying an effect on a beam of electrons.

Stern and Gerlach didn't interpret their experiment as being due to the spin of the electron, and it took another five years (and the later discovery by Goudsmit and Uhlenbeck) for this connection to be made. We now understand that the splitting of the beam of silver atoms into two shows that the spin of the electron can only take two possible values. Electrons either spin one way or the other, but no other possibilities are allowed. These two possibilities are often termed *spin up* and *spin down*, because the angular momentum of the electron is either up or down, parallel or antiparallel to the field gradient. In a sense, the Stern–Gerlach apparatus interrogated the electrons, asking them the question: what is your angular momentum along this direction? The answer seems to be 'up' from half of the electrons and 'down' from the other half. There is nothing special about the direction of the field gradient. You can orient a Stern–Gerlach experiment in other ways and yet you always get the answer 'up' from half of the electrons and 'down' from the other half. Moreover, there seems to be no way of predicting which way any individual electron will go when passing through the apparatus. On average, half of them go one way and half the other, but for any particular electron it is in the lap of the gods. Einstein famously said he thought quantum mechanics 'does not really bring us any closer to the secret of the "old one". I, at any rate, am convinced that He does not throw dice.' The Stern–Gerlach experiment would beg to differ and appears to give an example where you can see the dice being thrown.

The events in Germany in the 1930s had an effect on Stern and Gerlach that was rather reminiscent of their experiment: they split into two distinct and divergent trajectories. Stern emigrated to the US, became a US citizen in 1939 and served as a consultant to the war department during the Second World War. Though he resisted attacks on Jewish science, Gerlach remained in Germany and in 1944 became head of the German nuclear research programme, and was later one of the scientists detained at Farm Hall by the Allies.

Rotating spins

Mathematical physics at its best aims to follow in the wake of experimental discovery and put theoretical flesh on to the empirically discovered bones. The mathematical theory of spin began to be developed in the 1920s and some strange features soon emerged. We have been thinking about spin as if it is the self-rotation of a particle like an electron, much like a spinning cricket ball, but we have noted the strange feature that point particles can't really rotate. It was shown that though spin is a type of angular momentum, it represents a much more fundamental property of quantum mechanical wave functions. One of the first to understand this was Wolfgang Pauli who developed a theory of spin in 1927. Pauli developed a mathematical method for describing the electron spin, and this leads to some interesting consequences.

For example, take an electron spin and look at it, then rotate your head and look at it again. From the new perspective, the electron spin looks rotated. So what? Well, if you look at it from an angle of 360 degrees (i.e. you, the observer, must make a whole turn), then it looks like the negative of what you started with. That is just insane. If you turn something on its head it goes upside down, but if you turn it upside down *again*, surely you restore it to its original state, don't you? Not true with electron spin. You have to rotate it by 720 degrees to accomplish this. Two complete rotations are needed to restore the original state.

This idea sounds bizarre, but has been verified in experiments. In fact, there is an amusing parlour-game trick you can play to demonstrate this very effect in classical physics. Paul Dirac invented this and it is often known as his scissors trick (he

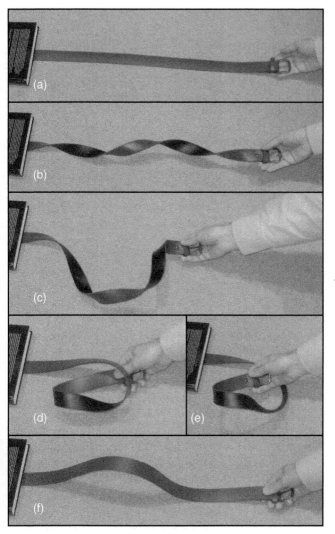

10. Dirac's scissor trick is more easily demonstrated with a belt with one end held fixed by, for example, placing it under a book

performed it by threading string around the eyes of a pair of scissors and a nearby chair), but in fact the effect can be demonstrated much more easily using a scarf or a belt, with one end held fixed by e.g. placing it under a book, as shown in Figure 10(a). The free end is rotated by two full turns (i.e. by 720 degrees) in the same sense and looks very twisted, as shown in Figure 10(b). It can be untwisted without rotating the free end in the opposite sense, simply by passing the free end around the middle of the belt and pulling taut, see Figure 10(c)–(f). Try it!

The Dirac equation

Pauli's theory of spin didn't include special relativity and so was never going to be the full picture. In 1928, Paul Dirac came up with a brilliant way of using special relativity and quantum mechanics to write down an equation describing the electron. Paul Adrien Maurice Dirac had been brought up in Bristol by his English mother and Swiss father. His father had insisted that only French was spoken at the dinner table, a stipulation that left Dirac with something of a distaste for speaking at all. After engineering and mathematics degrees at Bristol, Dirac moved to Cambridge to pursue doctoral research. His 1926 PhD thesis was entitled simply *Quantum Mechanics*. Following his discovery of what we now call the 'Dirac equation' and his other contributions, Dirac shared the 1933 Nobel Prize with Schrödinger 'for the discovery of new productive forms of atomic theory'. Following a sabbatical visit to work with the physicist Eugene Wigner at Princeton, Dirac married Wigner's sister Margrit in 1937. He usually referred to his wife simply as 'Wigner's sister' since, in his physics-dominated worldview, this description signified where she was located. Dirac had a very high view of mathematics, stating in the preface to his 1930 textbook that it was 'the tool specially suited for dealing with abstract concepts of any kind and there is no limit to its power in this field'. Later, he remarked that in science 'one tries to tell people, in such a way as to be understood by everyone, something that no one ever knew before. But in poetry, it's the exact opposite.'

Clarity for Dirac was fundamental, as was beauty, as it was 'more important to have beauty in one's equations than to have them fit experiment'. Failure to match the results of experimental data can be rectified by further experiment, or by the sorting out of some minor feature not taken into account that subsequent theoretical development will resolve; but for Dirac, an ugly theory could never be right.

Dirac spent a lot of time trying to find the right way to write down his new equation. Niels Bohr had asked Dirac in 1927 what he was working on. Dirac replied: 'I'm trying to get a relativistic theory of the electron.' Bohr replied that this problem had already been solved by another physicist (Oscar Klein). Dirac knew of this work but was aware of its flaws. His approach was quite different. 'A great deal of my work is just playing with equations and seeing what they give.' This playing led to a new equation which is staggering in its beauty and showed that spin is a natural consequence of the Dirac equation of an electron. It also yielded effortlessly the correct factor of two that Thomas had laboured hard to derive, as well as explaining the behaviour of an electron in a magnetic field (and another pesky factor, known as the g-factor, which came out right in the Dirac equation).

Dirac began his 1928 paper by stating that 'The new quantum mechanics, when applied to the problem of the structure of the atom with point-charge electrons, does not give results in agreement with experiment.' He explained how Goudsmit and Uhlenbeck's idea of spin had been shoehorned into quantum mechanics by Pauli and C. G. Darwin (grandson of Charles Darwin of evolution fame). However, he added that the 'question remains as to why Nature should have chosen this particular model for the electron instead of being satisfied with the point-charge'. Triumphantly, he was able to proclaim: 'It appears that the simplest Hamiltonian for a point-charge electron satisfying the requirements of both relativity and the general transformation theory leads to an explanation of all duplexity phenomena without further assumption.' This is as close as the taciturn Dirac gets to playing the giddy kitten. But his work has left other physicists simply open-mouthed in admiration. One of the pioneers of quantum electrodynamics, the Nobel Laureate Sin-itiro Tomonaga, writing on the Dirac equation, stated: 'We mortals are left reeling by this staggering outpouring of ideas from Dirac.'

By fusing relativity and quantum mechanics, the two great products of the early 20th-century revolution in physics, Dirac had assembled an equation which describes the origin of spin, the fundamental element of the magnetism of the electron. As the 20th-century wore on, magnetism would start bringing about its own quiet revolution. The upheaval this time would not be in theoretical physics but in consumer electronics, and it would bring about an extraordinary transformation in the way we store information.

Chapter 8
The magnetic library

Modern society is based around the storage and retrieval of extraordinary quantities of information. To see the problem we are up against, consider the following. It has been estimated that the text contained in all the books in the United States Library of Congress could be stored in about ten terabytes (a terabyte is 10^{12} bytes, or a million megabytes). If you were able to type out all the words spoken throughout their lifetimes by the hundred billion or so human beings who have ever lived, then the storage requirement for that text would be several exabytes (an exabyte is 10^{18} bytes, or a million million megabytes). In this context, consider that the human race accumulates currently many tens of exabytes of information per year. After a bit of thought, it is not hard to see why. Although raw text can be stored quite efficiently (the text from this book can be stored in well under a megabyte), pictures, audio, and particularly video are far more hungry for data storage. For example, a DVD-quality movie requires several gigabytes of storage, several thousand times more than needed to store the text from this book. Thus the modern data consumption of humanity, with many million digital cameras in circulation and actively used, not to mention mobile phones recording video clips, puts an immense demand on storage technologies that outstrips what the libraries of previous centuries had to deal with.

When computers first became available to the general public, information storage was bulky, slow, and expensive, and its capacity was tiny. Now it is compact, cheap, and its capacity is enormous. An individual human being now can own numerous CDs and DVDs, and store the data from these, and their cameras, plus any downloaded content, all on their computer's hard disk. That hard disk now has a capacity which is a sizeable fraction of that needed to store the text from all the books in the Library of Congress. This is nothing short of a data revolution and it has come about because of magnetism.

Sound beginnings

Long before the hard disk was used to store digital 1s and 0s, magnetism was being used in the emerging technology to record and transmit audio signals. The first moving-coil loudspeaker was designed by Oliver Lodge in 1898, though he never made one himself. It was only realized practically by Chester Rice and Edward Kellogg in the USA in the early 1920s. They used an electromagnet to provide a magnetic field and then wound a coil of wire around the base of a cardboard cone which was mounted in a non-magnetic metal frame. The audio signal was applied to the coil of wire and the presence of a current in the magnetic field led to a force on the cardboard cone which then vibrated. It was the vibrating cardboard cone which radiated the sound. In modern designs, the electromagnet at the base of the cone is replaced by a permanent magnet, and developments of more powerful, lightweight magnets have led to loudspeakers and headphones becoming less bulky and heavy.

Early microphones used a similar principle, only in reverse, with sound waves causing a vibration of the coil and inducing a voltage in it as it moves in a magnetic field. Most modern microphones, however, use other, non-magnetic technologies. A magnetic circuit remains at the heart of one sound-recording device: the pickup of an electric guitar. Developed first in the 1930s, an electric guitar pickup consists of a permanent magnet wrapped with many turns of fine wire. The vibration of the moving guitar string nearby induces a voltage in the coil and produces an alternating signal which can then be fed into an amplifier. A coil is, however, also prone to act as an antenna and therefore to pick up unwanted stray signals. A way of reducing interference is with the 1950s design of a humbucker pickup, consisting of two coils which are wound in an opposite sense and magnets which are arranged with opposite polarities; string motion induces a current in both coils in the same direction because both the winding direction and magnet polarity have been reversed. The interference signals are cancelled since their addition into the signal depends only on the winding direction, not on the magnetic polarities. This design not only removes the detected signal from local radio stations but also hum from transformers in power supplies, hence the name: humbuckers.

Recording

Microphones and guitar pickups can convert vibrations into electrical signals, but to make a recording you have to find a way of storing the information in those signals. The key insight needed to achieve this with magnetism comes from the fact that a small magnet can be magnetized in various directions, and the direction is stored in the magnet as a record of the way in which it was magnetized. This idea was first appreciated by an American engineer, Oberlin Smith, who had visited Edison and seen his phonograph. In 1878, he came up with a proposal for a method of magnetic recording which involved a silk thread wound on a drum. The thread was to be impregnated with small clippings of iron wire. The wire could then be passed through the core of an electromagnet connected to a microphone and hence magnetized according to the pattern of sound waves received by the microphone. The wire would then pass onto a second drum. The wire on the second drum could then be rewound onto the first drum and, in playback mode, the wire would run past a coil of wire, with the magnetic signal in the wire inducing a voltage in the coil as it went past. Thus, the magnetically stored information could be played back. The invention was never built, and Smith only published his idea in 1888, in an American technical journal *The Electrical World*. The Danish engineer Valdemar Poulsen built a magnetic wire recorder in 1899, christening his device the 'Telegraphone' and demonstrating it at the 1900 World Exposition in Paris. While there, he recorded the voice of Emperor Franz Josef of Austria, producing what is the oldest surviving magnetic audio recording. For magnetic recording to be a competitive technology, the signals (both on recording and playback) needed to be amplified and so further progress had to wait for the development of vacuum tubes.

A vacuum tube consists of various electrodes sealed in a glass bulb which has been evacuated of air. Electrons are emitted from a heated electrode (the cathode) into the vacuum by a process called thermionic emission. They are attracted to a positively charged electrode (the anode) and so a current can flow. Current cannot flow in the reverse direction because the anode is not heated. John Ambrose Fleming in London built such a vacuum tube in 1904 and his diode was the first vacuum tube rectifier. Things really took off when Lee de Forest at Illinois added a third grid electrode to make what he called an Audion and later became known as a triode. Applying a small voltage to the third grid could be used to control the current flow from cathode to anode, and this made the triode a very good amplifier of signals. Amplification was now possible and magnetic recording could be developed in earnest.

Various magnetic recording systems had now been designed, but an important breakthrough happened in the 1920s almost by accident. Certain high-end cigarettes were decorated with a gold leaf band and, to make a cheaper alternative, Fritz Pfleumer in Dresden designed a technique for coating paper with a cheaper gold-coloured bronze layer. Pfleumer realized that his cigarette-paper manufacturing technique could be adapted to make a paper tape coated in magnetizable material, such as small iron particles, and this could be used to replace wire recording. His magnetic tape recorder worked well and, although the paper tore easily, editing such tape was child's play; you literally cut and paste, exactly the same method as used for handling movie film. By the mid-1930s, the German chemical company BASF had found a way to replace paper with cellulose acetate, and to improve performance the iron particles were replaced with magnetite, Fe_3O_4 (lodestone appears again!).

Throughout the rest of the 20th century, further improvements were made, with the development of Fe_2O_3 (ferric oxide, often sold under the name 'ferric') and CrO_2 coatings (chromium dioxide, often sold under the name 'chrome') and polyester or PVC tape. Reel-to-reel tape recorders were sold widely in the 1950s and 1960s and used $\frac{1}{4}$-inch tape. These were succeeded by cassette tapes in the 1970s, the most popular format having 0.15-inch wide tape. With the advent of home video recorders, a wider half-inch magnetic tape was used for the popular VHS format. Magnetic tape survives to this day for use in some high-density data-recording formats.

Recording sound was originally carried out using analogue techniques, meaning that the louder the audio signal, the greater the degree of magnetization encoded. Now most audio data, and all computer files, are recorded digitally using the binary system of 1s and 0s ('bits' of information) as formulated by Gottfried Leibniz at the start of the 18th century (though the concept had been variously developed in India, China, and Africa many centuries before). In digital magnetic recording, a zero is encoded by magnetizing a small grain on the tape in one direction, a one is encoded by magnetizing it in the opposite direction.

Another place to store magnetic information is on a disk. The disk rotates and a read/write head can be guided to the appropriate region of the disk, and in this way information can be stored and retrieved. Though they have now become obsolete, floppy disks were extremely useful for transferring small amounts of data around from computer to computer. The first floppy disks appeared in 1971, were 8 inches in diameter and stored nearly 80 kilobytes. The standard high-density $3\frac{1}{2}$ inch disk of the late 1980s could be used to store a far more useful 1.44 megabytes. Their role has now been largely superceded by USB sticks and CDs, neither of which are magnetic technologies. Magnetism is, however, used for the much higher-density storage used in hard disks.

Hard disks

In the second half of the 20th century, some physicists had become fascinated by the goal of making very thin layers of magnetic material, struggling to fabricate perfect specimens in which the thickness was reduced down to a single atomic layer. Preparing thin layers of magnetic material had an important scientific aim, observing how magnetism behaves when you reduce the number of dimensions of the system. If magnetism occurs because of exchange interactions between neighbouring atoms, what happens when you reduce the number of nearest neighbours? Once this problem began to be investigated, it became possible to construct sandwiches of magnetic layers and non-magnetic layers, and various different structures could be investigated.

In the late 1980s, a new effect was discovered in a sandwich structure consisting of alternating ferromagnetic and non-magnetic layers. When the electrical resistance of such a structure was measured, it was found that it was much lower when the magnetic moments of the two ferromagnetic layers were parallel than when they were antiparallel. By getting the thickness of the intervening layer just right, the two ferromagnetic layers could be made to prefer to be antiparallel. Application of a magnetic field could then line them up. The change of resistance induced by a magnetic field is called magnetoresistance and the effect was so large it was christened *giant magnetoresistance*. The effect was discovered independently by groups led by Albert Fert in Orsay, France, and by Peter Grünberg in Jülich, Germany. Their discovery has led, via commercialization by IBM, to a very sensitive read-head that is found in all hard disk drives. For their part in the discovery, Fert and Grünberg won the 2007 Nobel Prize in Physics.

How does giant magnetoresistance work? As we have seen, electrons exist in two different spin states and so a macroscopic electric current contains electrons of both spins. When electrons travel through a ferromagnet, they are scattered more or less easily depending on whether or not their spin is aligned with the magnetization in the ferromagnet. Greater scattering occurs when the electron spin is antiparallel to the magnetization. That means that the parallel electrons (let's call them spin up) can travel much more easily and travel a path of low resistance. For spin-down electrons, it's much more like they are wading through treacle. If the two ferromagnetic layers are aligned in parallel, as shown in Figure 11, then spin-up electrons will pass through relatively unhindered, whereas spin-down electrons will struggle to get through. On the other hand, if the two ferromagnetic layers are aligned antiparallel, then both types of electrons will experience some scattering in one of the two ferromagnetic layers. In the first case (ferromagnetic layers aligned in parallel), there is an effective short circuit for spin-up electrons, so the total resistance is lower in that case. This is the origin of giant magnetoresistance.

This principle led to the development of the spin valve, a sandwich structure of layers which can be used as a sensitive detector of magnetic fields. In such a device, the spins in one ferromagnetic layer are held in a fixed orientation by situating this layer adjacent to an antiferromagnetic layer (which for complicated reasons serves to fix rigidly all the spins in the ferromagnetic layer along a particular direction). You then add a non-magnetic spacer layer and finally a second ferromagnetic layer which is free to rotate. This layer responds to the magnetic field close to the surface of the

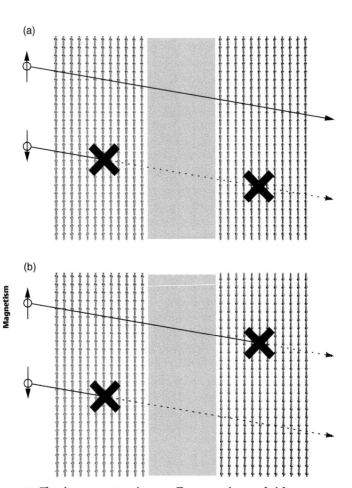

11. The giant magnetoresistance effect occurs in a sandwich structure. A different resistance is obtained when the two ferromagnetic layers are (a) parallel and (b) antiparallel. Spin up electrons short-circuit the device in case (a)

hard disk but the fixed layer does not. Passing a current through such a sandwich structure allows a current to flow relatively easily if the two ferromagnetic layers are parallel (valve open) or not if they are antiparallel (valve shut). The hard disk is spun underneath a 'read head' which contains a spin valve. The 1s and 0s encoded on the hard disk cause the free layer to switch back and forth, thereby opening and shutting the spin valve and giving rise to an electrical signal in the read head which is fed out of the hard disk unit and can be used by the computer. Spin valves were developed at IBM's Almaden labs in the late 1980s and early 1990s, and by the new millennium were in all hard disk drives.

In a conventional hard disk technology, the disk needs to be spun very fast, around 7,000 revolutions per minute. This means that the disk is moving under the read head at a speed which increases as the read head moves away from the axis of rotation but the instantaneous speed is typically around 20 metres per second (around the speed of a car on a motorway). The read head floats on a cushion of air about 15 nanometres (a nanometre is a millionth of a millimetre) above the surface of the rotating disk, reading bits off the disk at tens of megabytes per second. This is an extraordinary engineering achievement when you think about it. If you were to scale up a hard disk so that the disk is a few kilometres in diameter rather a few centimetres, then the read head would be around the size of the White House and would be floating over the surface of the disk on a cushion of air one millimetre thick (the diameter of the head of a pin) while the disk rotated below it at a speed of several million miles per hour (fast enough to go round the equator a couple of dozen times in a second). On this scale, the bits would be spaced a few centimetres apart around each track. Hard disk drives are remarkable.

Rotating the hard disk takes energy and of course a hard disk is a mechanical system which can wear out and fail. Although hard disks store an astonishing amount of information and are cheap to manufacture, they are not fast information retrieval systems.

To access a particular piece of information involves moving the head and rotating the disk to a particular spot, taking perhaps a few milliseconds. This sounds quite rapid, but with processors buzzing away and performing operations every nanosecond or so, a few milliseconds is glacial in comparison. For this reason, modern computers often use solid state memory to store temporary information, reserving the hard disk for longer-term bulk storage. However, there is a trade-off between cost and performance. Flash memory is becoming popular (particular for the USB sticks people attach to their key rings), but these have a limited lifetime and become unusable after 10,000 writing operations.

Racetracks

New technologies are being developed all the time for information storage technologies, and a particularly ingenious one is being developed by Stuart Parkin at IBM Almaden and is called racetrack memory. It contains no moving parts, offering greater reliability, and promises to be much faster than a hard disk and require less electrical power. The principle is simple. Bits of information are stored on magnetic nanowires, tiny ferromagnetic wires a few nanometres thick. Special components situated on a silicon wafer write the information onto the wires, and then shift the bits backwards and forwards along the wires. The bits of data zoom around the wire tracks like tiny nanoscopic racing cars, and the bits return to the reading device only when they are needed. The bits can be moved extremely rapidly, allowing very fast data access. A final device would require thousands of these nanowires on a single chip.

Between each one and zero on the racetrack is a domain wall, the thin region where the spins rotate round from up to down. The key to the racetrack memory is that if you apply a current along the nanowire, you fire electrons at the domains walls. Coming from a region storing '1', there are more electrons aligned in the up-direction than the down-direction but as they enter the region storing '0' some of the electrons have to flip spin. Spin is a type of angular momentum and because angular momentum must be conserved, the flow of electrons exerts a twisting force on the spins in the domain wall, causing the wall to slide along the wire. Exactly the same thing results from electrons coming from a region storing '0', except that here there are more electrons aligned in the down direction and some of these flip the other way as they pass through the domain wall; if you follow through the argument, you still find the domain wall slides in the same direction. Thus, simply by using a current, you can get the whole queue of domain walls to march in step, up the nanowire (or, by reversing the current direction, down the nanowire).

Money and timing

There are numerous creative, brilliant, and orignal ideas which could revolutionize the way we store information. But will they? What chance have they got to make it into the next generation of gadgets? The answer to that question depends on both money and timing. Money comes into it because, to displace an existing data storage technology, you need your new idea not only to provide improved performance or novel functionality but also to be manufactured at substantially lower cost, otherwise no-one will switch. Good timing is needed because some ideas work well at particular moments in history but not before their time (when other bits of technology are not ready) or after their time (when they end up being supplanted by something else). The window of opportunity may not be open for long and sometimes doesn't open at all.

A good example of this is provided by perpendicular magnetic recording, in which information is stored in bits which are magnetized up and down, in a direction perpendicular to the disk. It was known for many years that this was a more efficient way to perform magnetic recording (more bits can be stored per unit area) than the existing technology, namely parallel magnetic recording, in which the bits are magnetized left and right in the plane of the disk. However, it took many years to achieve this change because parallel recording worked well, the manufacturing techniques were well developed and in place, and the change to a new technology, involving retooling the manufacturing plants, had an associated cost. It was only when it became clear that the marketable advantages in performance of perpendicular recording outweighed the cost incurred by redesigning the manufacturing processes that the change happened.

Magnetic bubble memory, a concept pioneered in the 1960s at Bell Labs, is an example of a brilliant idea which never quite made it. A sheet of ferromagnetic material could be divided up into different domains of regions magnetized one way or the other. In certain materials, the spins like to be perpendicular to the plane and domains form as cylindrically shaped 'bubbles' which can be easily moved through the film. In a bubble memory, the ferromagnetic sheet stays stationary but the bubbles encoding the information are driven through the material. The concept was developed intensively but never made it beyond niche applications (it did prove to be possible to make very robust memories which were used in military applications). Bubble memory was overtaken by developments in hard disk technologies in the late 1970s.

Magnetic bits

In general, there is a strong economic drive to store more and more information in a smaller and smaller space, and hence a need to find a way to make smaller and smaller bits. In much the same way that the number of transistors that can be placed on an integrated circuit has roughly doubled every two years (following an empirical relation known as Moore's law), so the density of information of storage per unit area of hard disk has followed a comparable exponential increase, see Figure 12. In this most

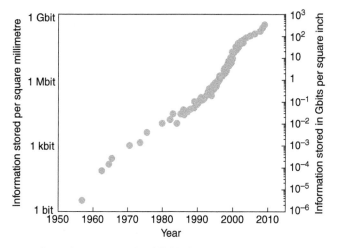

12. **Information storage on hard disks, shown as the number of bits stored per square millimetre (roughly the area of the head of a pin). The right-hand axis shows the information in the industry standard unit of Gbits per square inch. 1 kbit is one thousand bits, 1 Mbit is one million bits, 1 Gbit is one billion bits**

modern of industries, areal density (information stored per unit area) is conventionally quoted in units of Gbit/in² (billion bits per square inch). The figure also shows these quantities using a scale measured in bits per square millimetre. A square millimetre is about the area of the head of a pin. If we could read a trillion bits (1 Tbit or 1000 Gbits) per square millimetre, which equates to a bit per square nanometre, then we would be storing information at the atomic level. Current technology, at the time of writing, is some way off this limit. However, if the current rate of progress continues, we could be approaching it in the 2020s.

This progress towards greater miniturization comes at a price. The point is the following: when you try to store a bit of information in a magnetic medium, an important constraint on the usefulness of the technology is how long the information will last for. Almost always the information is being stored at room temperature and so needs to be robust to the ever present random jiggling effects produced by temperature (these are called *thermal fluctuations*). It turns out that the crucial parameter controlling this robustness is the ratio of the energy needed to reverse the bit of information (in other words, the energy required to change the magnetization from one direction to the reverse direction) to a characteristic energy associated with room temperature (an energy which is, expressed in electrical units, approximately one-fortieth of a Volt). So if the energy to flip a magnetic bit is very large, the information can persist for thousands of years (and information about the historical magnetic field of the Earth has been faithfully recorded in rocks for longer, see Chapter 9); while if it is very small, the information might only last for a small fraction of a second (clearly useless for a technological application). This energy is proportional to the volume of the magnetic bit, and so one immediately sees a problem with making bits smaller and smaller: though you can store bits of information at higher density, there is a very real possibility that the information might be very rapidly scrambled by thermal fluctuations. This motivates the search for materials in which it is very hard to flip the magnetization from one state to the other.

The ultimate goal for shrinking magnetic recording technology to its physical limit is to develop a technology that works at the molecular level. Some of the elements necessary for doing this may already be in place, but a realistic molecular-scale technology seems some way off. Synthetic chemists have developed a new type of storage medium which is called a *single molecule magnet*. These materials consist of assemblies of molecules, each one of which is a small cluster of metal ions surrounded by non-magnetic chemical groups (see Figure 13). In each molecule, the metal ions couple together to produce a giant spin in which some information can be encoded. Each molecule is only a nanometre or so across, so that in principle information could be stored at exceptionally high

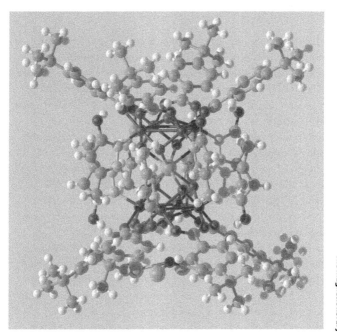

13. The molecular structure of a single molecule magnet. Is this the chemically engineered bit of the future?

densities. So far, though, it has not yet proved possible to find a way to address individual molecules at these densities. However, these molecular magnets have certain advantages over other possible approaches. First, in the conventional manufacture of small magnetic particles, one inevitably obtains a distribution of slightly different sizes. For the single-molecule magnets, because they are synthesized chemically, a set of completely identical molecules can be prepared. Second, molecular magnets seem to hold promise as storage systems for quantum information owing to their particular arrangement of energy levels and weak coupling to their nearby environment through the surrounding chemical groups.

Spintronics

Magnetism has a role to play not only because of information
storage but in the rest of electronics as well. In conventional
electronics, one only worries about the movement of electrons and
their associated charge. Now some scientists are beginning to
wonder if they can also harness the electron spin. This means that
in any circuit you can consider the flow of both spin-up and
spin-down electrons and using spin valves, spin-injectors, and
other spin-polarized circuit elements you can control and
manipulate these current flows separately. By marrying
conventional semiconductors like silicon with ferromagnets, and
using lithography and microfabrication techniques, you can
incorporate these different materials into new devices. This new
field has been christened 'spin electronics', or 'spintronics' for
short. It has already led to the development of spin transistors,
spintronic solar cells, domain-wall logic elements, and magnetic
random access memory (MRAM). Though it remains to be seen
which of these technologies will prove to be really useful, it seems
certain that the opportunity of using the spin of the electron has
given scientists working in this field a much-needed additional
dimension to think afresh about circuits and devices and a new
angle to bring to the invention of new technologies.

Chapter 9
Magnetism on Earth and in space

As Gilbert realized, the Earth is a giant magnet. Our planet produces a magnetic field which is typically 50 millionths of a tesla (the tesla is the unit of magnetic field, named after Nikola Tesla whom we met in Chapter 3). In this chapter, we will consider why the Earth behaves that way, how the Earth's magnetism protects us from lethal danger lurking in space, and also describe the magnetism in various other bodies in the Solar System and further out in the Universe.

Animals and their magnetism

First, we start with the Earth. The magnetic field produced by our own blue planet provides a direction which allows sailors to navigate the oceans. However, it's not only humans that can use the magnetic field of the Earth to find their way around. Many animals appear to be able to do it too. Turtles, bats, flies, newts, and lobsters all show evidence of using magnetic navigation, as of course do many migratory birds. A particularly impressive example of sophisticated magnetic navigation is that of the bar-tailed godwit. This remarkable bird makes a direct, non-stop flight from Alaska to New Zealand, travelling around 10,000 kilometres over the Pacific Ocean. During its heroic week-long journey, it flies only across ocean without passing over landmasses which could provide navigational markers. Furthermore,

New Zealand is a fairly small target which can be missed if the initial trajectory is out by a few degrees.

However, it is not entirely clear how magnetic navigation in animals works. Many animals are sensitive to electric fields, and it is possible that for some sharks and rays that move through sea water (an electrically conducting fluid), their passage through the Earth's magnetic field, or even the shaking of the shark's head from side to side, can induce a voltage (via Faraday's induction effect) that can be detected by electroreceptors. Even if this is the mechanism used by these fish, it won't work for animals that live out of the water. Here the mechanism might be connected with small crystals of magnetite (lodestone again!) which have been detected inside many animals, such as pigeons, honeybees, sea turtles, rainbow trout, and salmon.

Magnetite is certainly important in certain bacteria (called magnetotactic bacteria) which contain inside them chains of very small single crystals of magnetite. These chains line up with the Earth's magnetic field and the bacteria themselves are then also aligned, unable to rotate away from this fixed direction but confined to travel up and down a magnetic field line. This allows them to navigate deeper into the less oxygenated mud that they prefer. It is thought that the growth of crystals of magnetite in small cellular organisms began to occur in the period of the Earth's history when the oxygen content greatly increased, readily oxidizing iron and accidentally becoming incorporated in living tissue as part of iron uptake. Growth of such mineral structures inside living organisms is called biomineralization (other examples of this process of mineral growth inside animals include the formation of shells and bone).

Though more complex animals contain magnetite, it is not clear how they provide navigational function. In the beaks of homing pigeons, there are rather complex structures comprising magnetite (Fe_3O_4) crystals and maghemite (Fe_2O_3) platelets, but it is not known how they actually allow pigeons to find their way. It has been suggested that the movement of a small crystal might create an electric potential on a neuron or open an ion channel in a cell wall, but the details are sketchy and the problem unresolved.

An alternative mechanism has been proposed in which the Earth's magnetic field controls a particular chemical reaction between two free radicals (a free radical is an uncharged atom or molecule with an unpaired electron) and such reactions are now known to be unusually sensitive to the direction of weak magnetic fields. A photoreceptive protein, cryptochrome, has been found in the eyes of magnetoreceptive birds and this protein forms radical pairs after excitation with light. Recent experiments on the cryptochrome-containing fruit fly *Drosophila* have shown that they are sensitive to magnetic fields, but mutants lacking cryptochrome do not have the ability. If the eyes of birds use the same mechanism, it is possible that the bar-tailed godwit might be able to see its way to New Zealand with some kind of on-board 'satnav' that superimposes a magnetic image directly on to its vision. Or maybe it uses a number of cues, from the magnetite in its beak, the cryptochrome in its eyes, weak chemical signals up its nostrils, and the position of the Sun and stars, all helping to give a fuller picture of where it is. Biologists are still trying to figure out how this all works, and the rest of us will forever remain impressed and astonished at the wonders of the animal kingdom.

Why is the Earth magnetic?

Though the Earth's magnetic field permits navigation using a compass, the compass needle does not point in exactly the same direction at every location on the globe. This effect is called magnetic declination (sometimes magnetic variation) and means that there is a small correction to apply to your compass in order to work out where true magnetic north is. The first map of declination was produced by Edmond Halley (of comet fame) in

1701, based on his observations aboard a Royal Navy ship commissioned for the first scientific survey of the geomagnetic field. Halley realized that the Earth's magnetic field wasn't static and immovable but was slowly shifting, and therefore wrote that those who used his chart should remember that it was based on observations from the year 1700 and 'there is a perpetual, though slow change in the variation almost everywhere, which will make it necessary in time to alter the whole system'. The change in the Earth's magnetic field over time is a fairly noticeable phenomenon. Every decade or so, compass needles in Africa are shifting by a degree, and the magnetic field overall on planet Earth is about 10% weaker than it was in the 19th century. The reason for this time dependence of the Earth's magnetic field is something we will return to later in the chapter.

Halley's chart became an invaluable tool for sailors, guiding Captain Cook on his various voyages. In 1707, four Royal Navy ships under the command of Admiral Sir Cloudesley Shovell were lost on the granite reefs of the Scilly Isles with the consequent loss of more than 1,400 sailors. Their navigation used what is called *dead reckoning*, determining the ship's position each day on the basis of previous days' measurements supplemented by a crude estimate of the distance travelled during a day by making an assessment of the ship's average speed, taking into account currents and leeway (a process that easily led to errors steadily accumulating during a long voyage). Subsequent analysis has indicated that they did not use Halley's corrections for magnetic declination. The disaster shocked politicians into establishing a Board of Longitude in 1714 charged with the task of encouraging innovators to find a method to determine longitude at sea.

The navigational compass was subject to another effect which caused it to give misleading information: the magnetic influence of the ship itself. Iron objects were increasingly used on ships, from nails to anchors, and through the development of iron-clad ships the deviation became particularly severe. In a thunderstorm, a lightning strike could produce a sudden current which magnetized various bits of the ship and caused the compass to go haywire, just when it was needed the most. Very often the binnacle (the case on a ship in which the navigational instruments were mounted) would be assembled using iron nails, and sometimes the very case in which the compass was mounted would be made of iron, so that the compass was in the worst possible place for it to operate without error. In the mid-19th century, John Gray's binnacle (containing adjustable compensating magnets), and later a device designed by William Thomson using two spherical compensating magnets (known as Kelvin's Balls), did much to avoid maritime disasters by allowing ships' compasses to give less compromised measurements.

To provide the data for the successors to Halley's charts required monitoring of the magnetic field at various locations. In the 19th century, Carl Friedrich Gauss developed an elaborate mathematical analysis procedure to understand the variation of the magnetic field on the Earth as measured in various *magnetic observatories* scattered around the globe. Gauss, together with fellow German scientists Wilhelm Weber and Alexander von Humboldt, had petitioned the British Admiralty to get the observatories extended across the British Empire, and their worldwide geomagnetic observatory network, the *Magnetische Verein* coordinated from Göttingen, was one of the first major scientific collaborations carried out on an international scale, a forerunner of modern enterprises such as CERN. One of the great successes of Gauss's approach was the ability to show that the field predominantly originated from the Earth itself, just as deduced by Gilbert. However, Gilbert had guessed incorrectly that the Earth was a giant lodestone. Although the field measured at the Earth's surface contains a contribution originating from magnetized rocks in the Earth's crust, the temperatures much below the crust greatly exceed the Curie temperature of those rocks and so any magnetism would be destroyed. Something else must be going on.

In the early part of the 20th century, it was proposed that the Earth's magnetic field was due to a *self-sustaining fluid dynamo*, a concept developed by Joseph Larmor. The idea is that there is a circulation of hot conducting fluid in the Earth's core driven by thermal effects. As the conducting fluid moves through the magnetic field of the Earth, electrical currents are generated. It is these currents that produce the magnetic field of the Earth. This explanation sounds a bit like a magic egg which produces a chicken that lays the egg out of which it has itself hatched. But the 'self-sustaining' nature of Larmor's dynamo means that energy is being continually fed in from sources of heat inside the Earth's core, and this keeps the whole process cooking.

In fact, Larmor's model had to be modified following Thomas Cowling, who in 1933 provided arguments that proved if the motion of fluid is symmetric about an axis then no dynamo can be maintained. It is now thought that the nickel-iron fluid moves in convection cells, small cycles in which hotter fluid rises and colder fluid falls, and that turbulence is also important. Another factor is the rotation of the Earth which has an effect on this fluid much as it does on our atmosphere. Air doesn't move from points of high to low pressure, it circulates around them in cyclones and anticyclones, all driven via the Coriolis force originating from the Earth's rotation about its axis. In much the same way, the rotation of the Earth drives sideways motion in the liquid core of the Earth. The combination of all these process results in extremely complex behaviour which is possibly chaotic. As we shall see, this leads to an important effect on the long-term stability of the Earth's field.

Here comes the Sun

Larmor realized that his dynamo model might not only apply to the Earth but also to the Sun and he tried to use it to explain sunspots. These dark patches on the Sun had been noticed by Chinese astronomers and were being regularly observed well before the birth of Christ. Using a telescope, Galileo was able to watch sunspots moving across the surface of the Sun, leading him to conclude that the Sun was itself rotating. The number of sunspots follows a periodic cycle with a period of approximately 11 years, as shown in Figure 14, though the cycle is far from regular. Going further back, there was a very quiet period in sunspot activity which was observed from around 1645 to 1710 when the number of sunspots remained close to zero. This is known as the Maunder Minimum and coincided with the 'Little Ice Age' with uncommonly cold winters in North America and Europe, though whether the two events were connected is still debated.

In 1852, Sir Edward Sabine, running the network of magnetic observatories, noticed that periods of excessive fluctuations and disturbance in the Earth's magnetic field correlated well with the

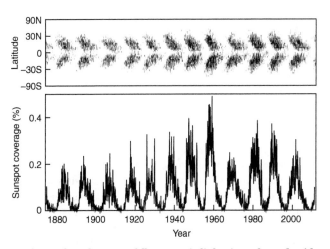

14. The number of sunspots follows a periodic but irregular cycle with a period of approximately eleven years. The top panel shows the 'butterfly diagram', describing how the sunspots appear in each cycle closer to the poles and then move in towards the equator

11-year cycle in the number of sunspots, showing that there is a contribution to the Earth's magnetic field from what is happening on the Sun.

The Sun is an extraordinary object. It contains more than 99% of the mass in the Solar System. The very high temperature (around a million degrees) in the outer atmosphere of the Sun, known as the solar corona, whips up protons, electrons, and alpha particles to speeds that exceed the escape velocity of the Sun. These particles stream out from Sun, forming what is known as the solar wind. This flow of charged particles generates a magnetic field, and this contributes to the interplanetary magnetic field. At the Earth, its strength is about 6 nanotesla.

The presence of the solar wind had been inferred from the observation that the tails of comets in the Solar System always point away from the Sun, regardless of the direction the comet is travelling. This signifies that a stream of particles flowing outward from the Sun is blowing comet tails away from the Sun. However, the flow of particles is not steady but varies with time because of what is happening on the surface of the Sun.

In 1908, George Ellery Hale, an American astronomer working at the Mount Wilson Observatory, found that the magnetic field at the dark centre of sunspots reached values as large as a few tenths of a tesla, a thousand times stronger than the field at the surface of the Earth. Hale did not need to visit the Sun to do this. He made these measurements without leaving Los Angeles county by collecting the light from particular regions of the Sun and splitting this light up into its constituent wavelengths. Within each spectrum, he could see particular spectral lines resulting from emission from certain atomic transitions and he looked for any splittings of these spectral lines resulting from the Zeeman effect (see Chapter 7) whereby a magnetic field causes spectral lines to split in energy by an amount proportional to the magnetic field. By repeating this experiment for light from every part of the solar disc, he was able to map out the magnetic field on the Sun. Hale's results showed that sunspots provide a marker for regions of large magnetic field on the Sun's surface.

The Sun contains gas which is ionized (the outer electrons are ripped off atoms leaving the remainder positively charged) so that the individual particles floating around have an electrical charge, and the resulting soup of ions and electrons forms a *plasma*. Plasma can be created on Earth, and some televisions use plasma in their displays. In plasma TVs, it is low-pressure xenon and neon gas confined in many tiny cells sandwiched between electrodes and glass, the electrodes triggering in turn for each cell and producing a current through the gas which leads to emission of light. This is all done at room temperature of course, but really interesting things can happen in a very hot plasma. Because the particles in plasma on the Sun are charged and moving very fast, they can produce, and interact with, magnetic field. In such plasmas, magnetic field lines can become trapped by regions of fluid and dragged along with the fluid as it flows around. These field lines can become twisted and tangled because of the complex motion of the fluid. Where field lines nearly cross, a phenomenon called magnetic reconnection can occur in which the field lines snap and recombine, releasing energy in the process and thus allowing a conversion between magnetic energy and kinetic energy. All of these processes are now known to be important in the Sun.

At the beginning of each sunspot cycle, it is found that the sunspots begin to appear in two bands, each at relatively high latitudes (high up in the northern and low down in the southern hemispheres), but towards the end of the cycle the two bands are found closer to the equator. This pattern is apparent in the so-called butterfly diagram shown in Figure 14. The explanation for this behaviour is not fully established. Suffice to say that understanding the solar dynamo requires use of the principles of fluid dynamics, plasma physics, and the interaction between lines of magnetic fields twisting and turning in a rotating, bubbling cauldron of convecting turbulent fluid.

Space weather

As the solar wind blows out from the Sun, it interacts with the Earth's magnetic field, producing something a bit like a shock wave which encloses a region called the magnetosphere, as shown in Figure 15. The magnetosphere has a tear-drop shape, extending about ten Earth radii in the direction towards the Sun, and possibly a couple of hundred Earth radii in the direction away from the Sun. The magnetosphere forms a protective layer around the Earth, cocooning the planet and providing some protection from the harsh environment of the solar wind.

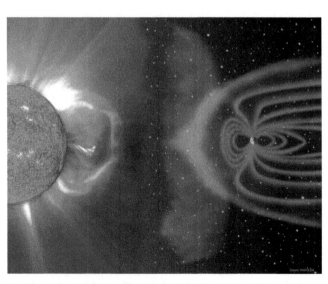

15. The magnetosphere of the Earth (right) interacts with charged particles produced in a solar storm on the Sun (left)

The lower region of the magnetosphere is known as the ionosphere, a series of concentric shells of electrons and ions existing in regions where the atmosphere is so thin that charged particles can survive for a reasonable duration without recombining. The lower layer is about 50 kilometres above the surface and contains ionized molecules such as nitrogen and nitrogen monoxide. The upper layers extend up to about 500 kilometres and contain ionized atoms. Ultraviolet radiation from the Sun is responsible for ionizing the various layers and so the density of ions depends on whether it is day or night. This is the reason why the propagation of short-wave radio depends on the time of day, because distant transmitting stations are picked up by radio waves bouncing off different layers of the ionosphere and so depends on the density of ions in these layers. Solar activity also affects the thickness and homogeneity of these layers and hence the quality of radio reception, as do meteor showers (which briefly increase ionization in the ionosphere).

The charged particles in the atmosphere interact with the Earth's magnetic field and spiral around the magnetic field lines, becoming concentrated around the poles. There the density of magnetic field lines is higher and collisions between particles occur, resulting in the emission of light. This produces a glow of coloured light in the sky, and close to the North Pole they are known as the northern lights, or aurora borealis (Boreas brought the north wind in Greek mythology). In the southern hemisphere, they are known as the aurora australis (the Latin word *australis* means 'southern').

Magnetic fields produce a force on charged particles which is at right angles to the magnetic field and to their motion. Consequently, charged particles execute corkscrew paths which wind around field lines. When a set of field lines bunch together (as they do near the poles), the particles can reflect from these regions and wind back the other way. As a result, many charged particles from the solar wind become trapped in doughnut-shaped regions around the Earth, known as the van Allen belts. Because the rotation axis of the Earth is tilted from the magnetic axis by around 11 degrees, the inner van Allen belt comes quite close to the Earth near the south Atlantic, producing an effect known as the *south Atlantic anomaly*. Satellites are particular susceptible to the effects of radiation in this region because the magnetic protection of the van Allen belts is at its weakest. Laptops on the space shuttle have been found to be more likely to crash when the shuttle passes over the anomaly due to the enhanced radiation.

Solar flares sometimes occur on the surface of the Sun. These are enormous explosions that heat gas to many millions of degrees, emitting X-rays and gamma rays as well as charged particles. In coronal mass ejections, bubbles of gas which are threaded with magnetic field lines are expelled from the Sun over a period of several hours. Sometimes these violent events can disturb the Earth's magnetosphere when the charged particles arrive at Earth and cause what is known as a geomagnetic storm. Occasionally, such storms can induce currents in long electrical power lines, potentially disrupting power supplies. Communication satellites are at risk from geomagnetic storms, first because of the increased cosmic radiation that directly impacts them and second because of the increased number and energy of charged particles that flow around them, inducing harmful voltages. It seems, therefore, that we should not only keep a weather eye on the Earth's atmosphere but also on the meterological conditions in the Sun.

Spacecraft flying at distance from the Earth are particularly vulnerable as they are away from the protection of the Earth's magnetosphere. A powerful solar flare in August 1972 occurred between the two last Apollo manned missions to the Moon. Had the flare struck during one of those missions, when the astronauts were outside the Earth's protection, the radiation dose received on board could well have been fatal. Major solar flares are relatively rare events, but are a worrying risk for future manned space missions, especially to more distant destinations such as Mars where the extended journey time increases the likelihood of a catastrophic event *en route*. This is a useful reminder that planet Earth not only provides humans with an atmosphere which we can breathe but also a magnetic shield which protects us from deadly cosmic radiation.

Field reversals

While navigating around a mountain pass, one can sometimes find that a magnetic compass deviates from the magnetic north pole due to the presence of locally magnetized rocks. This effect has been known for some time; in 1794, Alexander von Humboldt speculated the effect was due to lightning strikes of these rocks. (More recently, lightning has been suggested to be the agent responsible for magnetizing lodestone, since the Earth's magnetic field alone is insufficient to magnetize a lump of magnetite.) However, in the mid-19th century it was realized that volcanic rocks, on cooling through their Curie temperature, would become magnetized in the direction of the Earth's magnetic field at that time. Thus certain rocks contain a frozen-in fossilized record of the local magnetic field direction present as they were cooling.

There are some interesting complications in interpreting these fossilized records. Sometimes particular layers of rock have folded and buckled since they cooled and so the rocks may not be in the same orientation. Due to continental drift, the rocks may not even be in the same geographical location. (So, for instance, when locating the very ancient position of magnetic north, rocks on opposite sides of the Atlantic appear to give different answers, facts which are reconcilable only when one recalls that the Atlantic ocean is a relatively recent phenomenon, non-existent a couple of hundred million years ago, and so these rocks have drifted by different amounts.)

However, at the very end of the 19th century, well before continental drift had even been proposed, it was clear that at various times in Earth's history the magnetic field of the planet had reversed. From radiometric dating of various lavas, it is possible to reconstruct this history and it is shown in Figure 16. Magnetic north has been approximately at its current location for nearly 800,000 years, though its exact location has wandered around considerably in the northern hemisphere during that time and is currently marching north through the Canadian Artic at a rate of about 50 kilometres per year. (It also wobbles around about its average position by up to around 80 kilometres every day due to variable electric currents in the ionosphere and magnetosphere caused by the solar wind.) However, these historical data show that there are dramatic flips in the magnetic field during which magnetic north switches over to the southern hemisphere. The duration of a field reversal is thought to be relatively short, probably just a few thousand years. The phenomenon is not periodic, and the data in Figure 16 show that there are some rather long, relatively stable periods, like the one we are now in, and sometimes the field reversals follow more quickly in succession. The time interval between field reversals range from around ten thousand up to ten million years. On average, there seem to be three or four reversals every million years, again emphasizing that the current situation is relatively stable.

What causes these reversals and how might one be able to predict when they occur? The answer to the first question is not entirely settled, but is believed to be due to the chaotic nature of the Earth's

| 0 | 20 | 40 | 60 | 80 |

time (millions of years)

16. Magnetic field reversals over the last 80 million years. The black regions show the epochs in which the Earth's field has the same polarity as at present, the white regions where the polarity was opposite

dynamo and the complex inter-relation between convection in the liquid outer core, the flow of heat between that and the solid inner core (itself 70% as wide as the Moon and spinning very slightly faster than the Earth's crust), and twisting and tangling of magnetic field lines. This is an inherently non-linear problem and, like the cyclones and anticyclones in our atmosphere, one finds whirlpools in the conducting liquid outer core, driven by the Coriolis forces of the Earth's rotation. Just as for our own weather, the processes are complex and impossible to predict over a long period. Even with the most advanced supercomputers currently giving us meaningful models of how these field reversals might occur, it is not possible to say when our 800,000-year stable period will end. A reversal of the Earth's magnetic field may not be a comfortable phenomenon to experience: it is accompanied by a marked reduction in magnetic field as it flips, temporarily turning off our protection from space radiation and the solar wind.

The planets of the Solar System

Since the early 1960s, information about the magnetic fields around other planets and their satellites in the Solar System has been obtained from various missions, such as the measurements by the *Pioneer*, *Mariner*, and *Voyager* space probes which were fitted with on-board magnetometers. We now know that the magnetic fields from both Jupiter and Saturn are very large. Charged particles from the solar wind are trapped in their magnetic field lines and radiate electromagnetic waves. These produce radio emission detectable on Earth, which is modulated by the rotation of the planets, and was how the magnetic field from both planets was first inferred. In general, it has been found that the total planetary magnetism (or to use the technical term, its magnetic moment) is approximately proportional to the angular momentum of the planet, suggesting a similar physical model to that of the Earth: a geodynamo, with the rotational kinetic energy driving the magnetic field. Both the Moon and Mars have much weaker magnetic fields, probably due to their lack of a liquid core.

The large magnetic moments of Saturn and particularly Jupiter (which has a magnetic moment about 20,000 times that of the Earth) reflect the thick layer of hydrogen (under such strong gravitational pressure that it becomes metallic) around the core which supports active dynamos. In the case of Jupiter, the core may have a radius up to 75% of the planet's radius, and the magnetic field at the surface is about 0.4 thousandths of a tesla, around 10 times that on Earth. The Jovian magnetosphere extends out to several million kilometres in the direction of the Sun, and even further in the opposite direction, almost as far as the orbit of Saturn. If we could see it from Earth, it would appear as big as the Moon, even though Jupiter itself is only a bright dot in the sky.

Deep space

One of the most intense sources of magnetic fields in the Universe is found in neutron stars. The existence of these objects was suggested in the 1930s by Walter Baade and Fritz Zwicky who proposed that a dense object composed only of neutrons might be formed following the explosion of a supernova. They were not observed until 1967, when a research student in Cambridge, Jocelyn Bell Burnell, discovered objects in the sky that gave out periodic pulses of radio emission. To find a periodic signal was highly unusual and she originally termed her discovery LGM, which stood for 'Little Green Men'. However, it was soon realized that the signals she had discovered originated from rapidly rotating neutron stars (now known as pulsars). These objects emit electromagnetic radiation in beams aligned along the magnetic axis, which is typically at an angle to the rotation axis. It is only at the instant when the beam points towards the Earth that we can observe them as a brief pulse of electromagnetic radiation. The rotation of a pulsar is found to be very rapid, with orbital periods ranging from just over a millisecond to several seconds, the faster rotation rates being found for recently formed pulsars with the rotation reducing slowly over time.

Neutron stars are very dense, with the mass of a star squeezed into a sphere with a radius of only a few kilometres. During their formation, the typical stellar magnetic fields are compressed as the star collapses inwards, forcing the field lines together and magnifying the field strength so that fields of a hundred million tesla are created. In a particular type of neutron star called a magnetar, it is possible that the field could reach ten billion teslas. Magnetars are thought to form under particular conditions which cause an additional dynamo mechanism to wind the neutron star magnetic field up even further than would happen normally (though nothing about neutron stars deserves the qualifier 'normally'). Magnetars are somewhat unstable objects, and quakes on their surface trigger enormous releases of X-rays and gamma rays.

Very tiny magnetic fields also pervade galaxies, including our own Milky Way, and are typically a few billionths of a tesla. Where this field comes from is not fully understood, but the most plausible explanation is that an even tinier primordial 'seed' magnetic field, that was present in the early Universe, has been amplified by dynamo processes in the each galaxy. Understanding these weak interstellar fields may yield a clue to the formation of galaxies.

Fusion on Earth

Since the Second World War, there has been a huge research effort aimed at trying to achieve fusion reactions on Earth. In a fusion reaction, light nuclei are fused together to create heavier nuclei, releasing large quantities of energy in the process (and should be contrasted with fission, the splitting of very heavy nuclei into smaller parts, which is used in conventional nuclear power plants). Fusion is the process that keeps the Sun shining, and so obviously works. It is extraordinarily effective as a way of producing energy, and if we could build working fusion reactors we could solve the Earth's looming energy crisis. The fuel is cheap and plentiful (you just need a relatively small quantity of deuterium which can be extracted from sea water, and the available reserves can keep us going for millenia) and the technology is clean (the waste product is a very small quantity of helium and there are no greenhouse gases produced). The catch is that to get a fusion reaction to work you have to heat the plasma to two hundred million degrees Celsius. That is a temperature which is of a quite different order to anything normally encountered; the hottest furnaces rarely exceed a few thousand degrees Celsius.

In the Sun, those sorts of temperatures are achieved quite naturally, but on Earth it's quite a different matter. If you put the plasma in a container and start to heat it up, the container will vaporize at a few thousand degrees Celsius. How can you possibly contain the plasma at two hundred million degrees Celsius? Once again, magnetism provides an answer. If the plasma is driven around a circular path it produces a magnetic field which tends to confine it into that path. If additional magnets are deployed in various directions, the plasma can be carefully controlled and confined, although the process is rather complex as plasma is a slippery customer and keeping it in a well-defined doughnut shape is rather like holding on to a particularly aggressive snake. Fusion has been achieved in the experimental fusion reactor in Culham, Oxfordshire, albeit only for a matter of seconds. It has only been possible to approach the break-even point, at which more energy is produced than is used to start the whole thing going in the first place (and clearly you have to exceed that point by some margin for fusion to be remotely useful for practical power production). The International Thermonuclear Experimental Reactor (ITER, see Figure 17), is currently being built in Cadarache, in the south of France, and has been designed to produce 500 MW of output power for only 50 MW of input power. It uses large superconducting magnets capable of producing more than 10 tesla of magnetic field to confine the plasma and stop it touching the walls of the vacuum chamber. Significant technical problems to overcome include the degradation of the superconducting magnets by the bombardment of neutrons that are produced in

17. ITER, the International Thermonuclear Experimental Reactor

the hot plasma. Building fusion reactors requires the combination of a wide range of skills from different fields, from heavy engineering and nuclear physics to materials science.

But fusion research is a long haul. ITER probably won't be fully operating until the 2020s. Plans are afoot for ITER's successor which aims to put power in the grid by 2040, so that fusion might become a practical reality in the second half of the 21st century. However, whenever a fusion power plant eventually comes to a district near you, there's a fair betting that it will involve the use of a big magnet.

Chapter 10
Exotic magnetism

The magnetism of lodestone is a state of parallel alignment in which all the spins within it line up. However, the zoo of magnetism also houses some far stranger animals in which spins are arranged in more complicated and outlandish configurations. This final chapter describes some examples of exotic magnetism and illustrates a few of the surprising and complex ways in which atomic magnets interact in solids.

Antiferromagnetism

Depending on the manner in which atoms in a magnetic solid interact, it can be possible that the spins do not line up all in parallel. If, in a spirit of perverse contrariness, the spins of each magnetic atom prefer to do the opposite of their neighbours, then one ends up with a system as shown in Figure 9(b) in Chapter 6: an antiferromagnet.

Lots of compounds, in particular oxides, are antiferromagnets. This is due to a feature of the magnetic interaction between two magnetic species in which an oxygen ion 'gets in the way', and which results in the spins on the magnetic species being forced to align antiparallel with each other. One special example of this is a copper oxide also containing lanthanum (its precise chemical formula is La_2CuO_4). This compound contains layers consisting of squares with copper ions on the corners and oxygen ions in between each copper ion. The compound is an antiferromagnet with copper spins ordering in an antiparallel arrangement. However, if you suck some electrons out of the layers (by chemically fiddling around with the lanthanum sitting between the layers), you can force the material to enter a superconducting state in which its electrical resistance completely vanishes. Antiferromagnetism seems to be mixed up with the riddle of high-temperature superconductivity, a hot topic in physics, and so the magnetic properties of this and related antiferromagnets remain of great interest.

Born to frustration

Ferromagnets and antiferromagnets are ordered magnets. Once you get these materials cold enough, the spins on each magnetic atom align all along the same direction and either line up all in parallel (ferromagnetism) or alternate in antiparallel (antiferromagnetism). But things get rather interesting when it is not possible to find an arrangement of spins that satisfies the interactions between them. A good example of this sort of problem is found in the so-called love triangle. The frustration comes in this sort of situation because when A loves B and B loves C, the relationship between A and C is inevitably frosty. An even more complicated problem occurs when three spins sit on the corners of a triangle when the exchange interactions between them are such that each spin wants to lie antiparallel to each of their neighbours (see Figure 18). There is no easy solution to this problem, and in classical models the spins have to adopt some kind of uneasy compromise where the pain is shared around.

The situation gets even more complicated when experimenters look for materials in which spins sit on the corners of a vast network of triangles, in so-called *kagome lattices* (named after a type of Japanese basket-weaving pattern), and in networks of corner-sharing tetrahedra (of which more later). The spins which

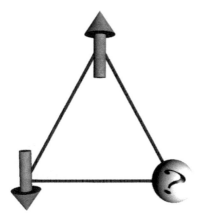

18. Two spins are placed on corners of a triangle. The antiferromagnetic interaction between them is satisfied. But how do you put a spin on the third corner? The situation is frustrated

populate these lattices are tortured by the frustration of not being able to find a state which satisfies all the exchange interactions.

Even cooling to low temperatures is not enough to force the spins to line up in any semblance of an ordered arrangement. Sometimes inherent randomness, either because of the locations of the spins inside the solid or the nature of their interactions prevents order occurring, but instead the spins slow down and begin to settle into some kind of random and disordered state characterized by dynamics on a long timescale. Such systems are called *spin glasses* to stress the resemblance of their magnetic state to the positional disorder of the atoms in amorphous solids (which are commonly called *glasses*).

If disorder isn't present but the interactions are tuned in such a way that magnetic order is frustrated, then another option is to form what is known as a spin liquid, a fluid-like magnetic state in which the constituent spins are highly correlated with one another but continue to fluctuate as the temperature is lowered towards absolute zero. The existence of a spin liquid was proposed by the American physicist Philip Anderson in the 1970s. He pictured spins on a lattice with antiferromagnetic interactions between nearest neighbours and postulated that pairs of spins would join together in the non-magnetic singlet state described in Chapter 6, a Schrödinger-cat-like combination of the (up down) and (down up) configurations.

The new twist is that Anderson realized that there was more than one way for the spins on a lattice to pair up into singlets. Playing by the rules of the weird world of quantum mechanics, he suggested that all possible pairing configurations exist simultaneously in a giant superposition which realizes all possible situations. The spin liquid state is like a giant dance floor, where a multitude of dancers get to tango with every other dancer, but all possible dancing pairs exist simultaneously! Experimental realizations of these strange spin liquids are now being found.

Finding new magnets

How are new magnetic materials found? There are a number of strategies that are frequently employed. One is to play with alloys, homogeneous mixtures of metallic elements and tinker with the balance of ingredients in order to optimise some desired property. Another option is to design some complex chemical structure using the techniques of solid-state chemistry, essentially employing a type of high-level cordon-bleu cookery in which the ingredients are more exotic, the ovens are hotter, and the cooking times longer.

One very rich area of current research in this area is to use complex molecular species to assemble molecule-based (rather than atom-based) magnetic materials. This approach takes a lead from Nature, which also employs molecular components to construct biological systems. The advantage of this strategy, well known to biochemists, is that small chemical adjustments to the molecular building-blocks can lead to extremely subtle changes in functionality, allowing some desired attribute of the final product to be carefully tuned. This route is leading to new types of magnetic material including some with interesting optical properties.

Another current hot topic is the discovery of new materials which combine interesting electric and magnetic properties. One family is termed multiferroics because of the combination of different types of 'ferro' order: ferromagnetism (the alignment of magnetic moments) and ferroelectricity (the alignment of electric dipoles), and possibly ferroelasticity (the alignment of elastic deformation). Though the prefix 'ferro' derives from the Latin *ferrum* for iron, it is now being used to describe the spontaneous alignment of a variety of physical properties in a manner that is reminiscent of the alignment of magnetic moments in iron. The combination of 'ferro'-orders in multiferroics often means that the different properties interact with each other. This then allows the possibility that you can switch magnetically ordered states using electric fields, or electrically ordered states using magnetic fields. The former case could be particularly useful since magnetically ordered states are good at storing information (see Chapter 8), but magnetic fields are more complicated to apply to control those states. On the other hand, electric voltages are easy to apply at a microscopic level, so an electrically controlled magnetic state could be extremely useful. Electrical control is achieved because the electric voltage switches the ferroelectric state which interacts with the magnetic state and causes it to switch too.

Spin ice and magnetic monopoles

A bar magnet contains a north pole at one end and a south pole at the other. If you chop the magnet in half, in the hope of isolating one of the poles, you end up forming a new south pole on the other end of the half with the north pole, and a new north pole on the other end of the half with the south pole. Poles come in pairs and even a single atom behaves as a magnetic dipole (meaning two poles, a north and a south). One of Maxwell's equations (see Chapter 4) enshrines the non-existence of isolated poles, also known as monopoles.

The line of reasoning which forbids the existence of monopoles has been questioned by various physicists, including Henri Poincaré and J. J. Thomson. It was Paul Dirac who came up with an argument that free magnetic monopoles should exist and that their magnetic charge would be quantized. However, he became disillusioned by the lack of any experimental evidence for the existence of magnetic monopoles, something which persists to this day. Some grand unified theories predict that magnetic monopoles may have been produced in the early Universe, and at the time of writing, there is speculation that evidence for magnetic monopoles might show up at the Large Hadron Collider.

However, very recently, objects *behaving like* magnetic monopoles have been found in a solid that you can hold in your hand. To explain this discovery, I must begin with a rather fascinating compound called dysprosium titanate. This material contains dysprosium, titanium, and oxygen, but the only atoms that we need to focus on are the dysprosiums because these are the magnetic ones. The structure of this compound is called a pyrochlore, because it is the same as the mineral pyrochlore found in certain rocks (and named pyrochlore, greek for 'green fire', because of the colour it turns when you place it in a hot flame). The dysprosium atoms sit on the corners of a tetrahedron and these tetrahedra all link together in a three-dimensional arrangement so that the corner of one tetrahedron touches the corner of another one. Because of an electrostatic interaction between the electrons on a dysprosium atom and those on some of the other atoms (the oxygens and titaniums that we've been ignoring), it turns out that the magnetic moments on the dysprosium atoms can either point into the centre of the tetrahedron, or out of it. Those are the only two options.

Furthermore, the magnetic interactions between the dysprosiums on the tetrahedra dictate that two of the magnetic moments can point in and two of them can point out. It doesn't matter which two are in, and which two are out, but the rule: 'two-in, two-out' has to be followed (see Figure 19(a)). When you extend this throughout the whole crystal, the freedom to choose which spins are pointing in and which are pointing out gives an additional entropy to the system – a residual disorder which persists to low temperature – and this can be measured in experiments.

When this was figured out in the 1990s, it was noted that exactly the same kind of behaviour had been deduced decades before in ice. Ice is of course frozen water, H_2O, and in ice the oxygen atoms are also arranged in a pyrochlore lattice, that is, a network of tetrahedra. However, each oxygen atom comes with a pair of hydrogens (because it's H_2O) and it turns out that this pair of hydrogens can either point in towards the centre of the tetrahedron or outwards in the opposite direction. For each tetrahedron, two of the oxygens have their hydrogens pointing in and the other two of them have them pointing out. These are the so-called *ice rules* that have to be obeyed. Because you can satisfy the ice rules in various ways (you have the freedom to choose which pairs of oxygens have their hydrogens pointing inwards), there is an extra contribution to the entropy. This residual disorder was also observed in experiments on ice, causing confusion until 1936, when Linus Pauling came up with the explanation I have just described. Because the essential physics of dysprosium titanate is analogous to that of ice (once you substitute 'dysprosium magnetic moment' for 'a pair of hydrogens'), this material was called a *spin ice*.

This may all be very interesting, but so far there has been no mention of magnetic monopoles. It took until 2007 for someone to work out what can happen to spin ice if you give it some energy and disturb it out of its equilibrium state. In other words, we know how spin ice will behave at very low temperature: all the

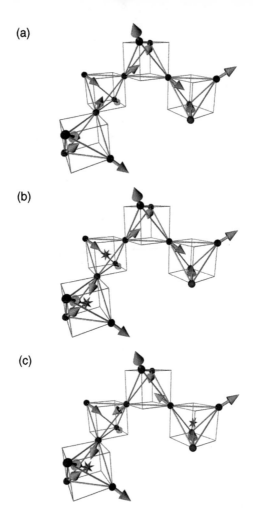

(a)

(b)

(c)

19. (a) In the ground state of spin ice, each tetrahedron has two spins pointing in, two spins pointing out. (b) Flipping one spin then wrecks the spin ice conditions for two adjacent tetrahedra (indicated by stars at the centre). The left one has three spins pointing in, one out; the right one has three spins pointing out, one pointing in. (c) By flipping spins between tetrahedra, the breaking of the spin ice conditions can be removed to a more distant tetrahedron

tetrahedra will obey the '2-in 2-out' rule and that is that. But what if we put a mistake into the structure? What if we reverse a single spin? Well, one of the tetrahedra will have '3-in 1-out' and, because the tetrahedra are corner-sharing, a neighbouring tetrahedron will have '1-in 3-out' (see Figure 19(b)). The key insight was to realise that this second tetrahedron can be restored to its proper '2-in 2-out' state by flipping a magnetic moment on its other side. What this does is to shift the '1-in 3-out' configuration along. What's more, we can repeat the trick and shift the '1-in 3-out' configuration further away from the '3-in 1-out' one (see Figure 19(c)), so that these two rule-breaking configurations can each move independently through spin ice.

Let's summarize what we've done. We started with perfect spin ice and we then flipped a single magnetic moment. This messed up two tetrahedra next to each other: one became '3-in 1-out' (let's call this +) and the other '1-in 3-out' (let's call this −). Then we realized that these two 'messed-up' tetrahedra can separate, and the + and − can each wander off through the crystal. What is rather amazing about this is that these two forms, the + and the −, behave like two separate magnetic monopoles of opposite charge.

This last supposition was initially made using theoretical calculations in which it was shown that the force between these two messed-up tetrahedra was exactly what you would expect if they were magnetic monopoles. Clever experiments soon followed that provided convincing evidence that these excitations can indeed be thought of as magnetic monopoles.

The emergent Universe

Are the magnetic monopoles in spin ice a sleight of hand? Partly. No-one is suggesting that cosmological monopoles have been found, nor that Maxwell's equations have been violated. Our current understanding of particle physics is almost certainly incomplete and there are very likely numerous surprises around the corner and the existence of magnetic monopoles might well be one of them. The magnetic monopoles in spin ice are *essentially* composed of spins, little atomic magnetic dipoles, each one obeying Maxwell's equations. However, the fact that their collective behaviour produces effects which are like magnetic monopoles is a non-trivial observation. The most efficient description of the phenomenon is by using a model of magnetic monopoles. When it comes down to it, that is what physics is all about: finding the most efficient, economical, and elegant descriptive model to account for the observations, and in the case of spin ice that is exactly what the magnetic monopole picture does for us.

In magnetism, we have some very well-articulated physical models which account for the interaction between spins. The properties that *emerge* from the complex balance of interactions can often be completely unexpected and although they arise from understood interactions, the complexity and many-atom nature of the physical problem under consideration leads to new behaviour. Even simple ferromagnetism is a phenomenon that does not occur in a single atom, but you need a multitude of atoms to see it. It is an example of an emergent phenomenon that does not admit to the simple reductionism (breaking things down to their smallest unit) that is often effective in many other branches of science.

Conclusion

Not only has magnetism changed our picture of the Universe, but it has also changed the actuality of our world. By giving us compasses, it allowed us to navigate the oceans, and by giving us motors, generators, and turbines, magnetism has given us plentiful power. It lies behind many of our electrical sensors, helps us in recording and playing music via microphones and loudspeakers, and has transformed the way we store information. Magnetism has played a crucial part in our maritime, industrial, and information revolutions.

The study of magnetism is all about building mathematical descriptive analogies which encode the complex and subtle interactions between units. Many of these models have been adapted subsequently for use in other spheres, for example in complexity theory, which is used to model biological and sociological processes. Cooling a lodestone through its Curie temperature induces a transition from disordered paramagnetism to ordered ferromagnetism, and understanding this has inspired work on other phase transitions, including those that are thought to occur in the very early Universe. Studies of frustrated magnetism have been used to inform other fields in which frustrated interactions impede the achievement of a state which satisfies all the constraints imposed upon it. And as we have just seen, magnetism has also added to our understanding of the emergent Universe, in which the collective behaviour of individual interacting units produces an effect which is incomprehensible from the study of a single unit, a new profound property *emerging* from the seething and bubbling interactions of the multitude.

But perhaps most of us all, magnetism has aroused humanity's basic curiosity. The image in Figure 1 of the pattern produced in iron filings from a magnet shows an experiment that can be done by a child. But that experiment illustrates relativity (magnetic fields are a relativistic correction of moving charges), quantum mechanics (the Bohr–van Leeuwen theorem forbids magnetism in classical systems), the mystery of spin (it is electron spin which produces the magnetism), exchange symmetry (which keeps the spins aligned), and emergent phenomena (many spins doing what a single spin cannot). With this in mind, one cannot escape the conclusion that magnetism itself is emblematic of the mystery, the wonder and the richness of the physical world.

Mathematical appendix

Maxwell's equations were given in non-mathematical form in Chapter 4. In this appendix they are written using vector notation and employing the vector differential operator ∇.

Maxwell's first equation is written

$$\nabla \cdot \mathbf{E} = \frac{\rho}{\epsilon_0}$$

where ρ is the charge density and ϵ_0 is a constant, known as the permittivity of free space. Maxwell's second equation is written

$$\nabla \cdot \mathbf{B} = 0.$$

Maxwell's third equation is written

$$\nabla \times \mathbf{E} = -\frac{\partial \mathbf{B}}{\partial t},$$

where the symbols $\partial/\partial t$ signify 'rate of change of'.

Maxwell's fourth equation is written

$$\nabla \times \mathbf{B} = \mu_0 \mathbf{J} + \mu_0 \epsilon_0 \frac{\partial \mathbf{E}}{\partial t},$$

where $\mathbf{J}$ is the current density and μ_0 is the permeability of free space. These equations are valid for free space and need to be modified in the presence of matter.

Further reading

Non-technical books

P. Fara, *Fatal Attraction* (New York, MJF Books: 2005).

A. Gurney, *Compass* (New York, W. W. Norton: 2004).

J. Hamilton, *Faraday* (London, HarperCollins: 2003).

F. A. J. L. James, *Michael Faraday: A Very Short Introduction* (Oxford, Oxford University Press: 2010).

Lucretius, *On the Nature of the Universe*, tr. R. Melville (Oxford, Oxford University Press: 1997).

H. W. Meyer, *A History of Electricity and Magnetism* (Norwalk, Connecticut, Burndy Library: 1972).

A. E. Moyer, *Joseph Henry* (Washington, Smithsonian Institution Press: 1997).

A. Pais, *Inward Bound* (Oxford, Oxford University Press: 1986).

S. Pumfrey, *Latitude and the Magnetic Earth* (Duxford, Icon: 2002).

C. A. Ronan and J. Needham, *The Shorter Science and Civilisation in China*, volume 3 (Cambridge, Cambridge University Press: 1986).

H. Schlesinger, *Battery* (New York, HarperCollins: 2010).

G. L. Verschuur, *Hidden Attraction* (Oxford, Oxford University Press: 1993).

J. B. Zirker, *Magnetic Universe* (Baltimore, John Hopkins University Press: 2009).

Specialist accounts

S. Blundell, *Magnetism in Condensed Matter* (Oxford, Oxford University Press: 2001).

S. Chikazumi, *Physics of Ferromagnetism* (Oxford, Oxford University Press: 1997).

J. M. D. Coey, *Magnetism and Magnetic Materials* (Cambridge, Cambridge University Press: 2010).

O. Darrigol, *Electrodynamics from Ampére to Einstein* (Oxford, Oxford University Press: 2005).

W. Lowrie, *Fundamentals of Geophysics*, 2nd edn. (Cambridge, Cambridge University Press: 2007).

D. C. Mattis, *The Theory of Magnetism Made Simple* (London, World Scientific: 2006).

James Clerk Maxwell, *A Treatise on Electricity and Magnetism* (Oxford, Clarendon Press: 1873).

N. Spaldin, *Magnetic Materials* (Cambridge, Cambridge University Press: 2011).

S. Tomonaga, *The Story of Spin* (Chicago, University of Chicago Press: 1974).

Index

Index

ONLINE CATALOGUE
A Very Short Introduction

Our online catalogue is designed to make it easy to find your ideal Very Short Introduction. View the entire collection by subject area, watch author videos, read sample chapters, and download reading guides.

http://fds.oup.com/www.oup.co.uk/general/vsi/index.html

SOCIAL MEDIA
Very Short Introduction

Join our community

www.oup.com/vsi

- Join us online at the official Very Short Introductions **Facebook** page.
- Access the thoughts and musings of our authors with our online **blog**.
- Sign up for our monthly **e-newsletter** to receive information on all new titles publishing that month.
- Browse the full range of Very Short Introductions online.
- Read **extracts** from the Introductions for free.
- Visit our library of **Reading Guides**. These guides, written by our expert authors will help you to question again, why you think what you think.
- If you are a teacher or lecturer you can order inspection copies quickly and simply via our website.